Antennas and Site Engineering for Mobile Radio Networks

For a listing of recent titles in the
Artech House Antenna Library,
turn to the back of this book.

Antennas and Site Engineering for Mobile Radio Networks

Bruno Delorme

ARTECH HOUSE

BOSTON | LONDON
artechhouse.com

Library of Congress Cataloging-in-Publication Data
A catalog record for this book is available from the U.S. Library of Congress.

British Library Cataloguing in Publication Data
A catalogue record for this book is available from the British Library.

ISBN-13: 978-1-60807-703-8

Cover design by Vicki Kane

© 2013 ARTECH HOUSE
685 Canton Street
Norwood, MA 02062

10 9 8 7 6 5 4 3 2 1

Contents

Foreword

In the last twenty years; that is, in a very short period of time as a matter of fact, mobile radio networks have emerged to become—whether we like it or not—an integral part of our lives. The complexity of mobile radio networks has evolved rapidly, and the engineers and technicians involved in their deployment and functioning require knowledge that is thorough, specialized, and sometimes unexpected, in various fields such as radio, computers, civil engineering, or marketing—not to mention sociology and medicine!

In the purely technical realm, regarding radio and more specifically antennas (which are a main and yet often neglected element of any radio network), two categories of works have existed to date. In English and in particular German literature, many scientific works were developing their concepts and their applications with much accuracy and many theoretical bases. On the other hand, a number of books with a clearly empirical purpose tried to explain—without always looking seriously into the theory—the functioning and manufacturing of antennas from the point of view of craftsmanship. It is interesting and curious to note that none of these books were specifically dedicated to antennas for mobile radio networks until now.

The great merit of Bruno Delorme is to have succeeded in doing, as it were, a triple split in this book; that is, in reconciling the theory and making it understandable for anyone with basic knowledge of mathematics and physics, detailing the practice of antennas, and at the same time focusing explicitly on the field of mobile radio networks.

If I am sincerely delighted at this initiative, it is not only because this book will serve as a support for all those persons, male and female, growing more and more numerous, who are professionally involved in defining and using antennas for mobile radio networks, but also because it will allow laypeople to take an interest in this subject and to find in it explanations based on many subjects so often discussed without full knowledge of the facts.

As far as I am concerned, it is clear that Bruno Delorme's book will be useful, to me as well as to my teams throughout the world, as a medium for the great amount of training we do throughout the year for the engineers and technicians working directly or indirectly for cellular operators.

It is thus with great pleasure that my collaborators and myself have modestly accompanied and supported Bruno Delorme in the realization of his project, which we confidently hope will quickly turn, thanks to its simplicity and clarity, into a reference book in this fascinating field!

Christian Harel
Managing Director of Kathrein France

Preface

The idea of writing this book in English occurred to me from observing that, currently, there are no books available on the industrial antennas used in the infrastructures of mobile radio networks, even though the number of Web sites on antennas has been increasing throughout the world over the last several years. According to one of the last censuses, there are 47,000 antenna sites in France, which correspond to a very considerable number of mobile radio networks.

It is necessary to know that these networks comprise not only the infrastructures of mobile phones managed by the cellular operators, but also all the private networks used by groups such as fire brigades, police forces, gendarmerie, railways, electric companies, gas companies, city halls, bus and tram companies, subway systems, taxi and ambulance companies, airports, harbors, and highway systems.

What Is Currently Available in the Bookstores?

- Theoretical books with very few industrial applications;
- Books on antennas, used since the beginning of broadcasting, for transmitting stations in long wave (LW), medium wave (MW), and shortwave (SW);
- Books on antennas used in the HF band (2–30 MHz) for long-distance links (intercontinental links) of audio or data transmission;
- Practical books for ham radio operators, essentially also in the HF band.

There is therefore a need to be able to find in bookstores a book devoted to:

- Basic overviews on antennas and their applications to industrial antennas, which are installed on pylons of mobile radio networks in the VHF and UHF bands (from 34 MHz–2.2 GHz) for mobiles, and in the SHF and EHF bands for microwave links.

- Engineering of antenna sites detailing the best way to install antennas (possibly multicoupling), and for the protection of receivers against interference frequencies created or induced on the sites.

This book, composed of three parts, thus meets these two expectations.

Part I: Antennas in Mobile Radio Networks

- Chapter 1: "Fundamentals of Antennas." How does an antenna radiate? Presentation of the characteristics of the electromagnetic wave. Definition of antenna parameters, radiation intensity, polarization, directivity, isotropic and relative gain, input impedance, bandwidth, and aperture angle at half power.

- Chapter 2: "Omnidirectional Vertical Wire Antennas." Calculation of the characteristics of vertical omnidirectional wire VHF and UHF antennas (gain, radiation patterns, aperture angle . . .) for the dipoles and antennas with multiple dipoles. Examples of industrial antennas.

- Chapter 3: "Panel and Yagi Antennas." Calculation of the characteristics of directional wire antennas VHF and UHF: panel antennas, YAGI antennas. Examples of industrial antennas.

Part II: Antennas Site Engineering

- Chapter 4: "Antenna Coupling." Coupling of antennas between them on a pylon or on a terrace.

- Chapter 5: "Antenna Coupling with Mast and Pylon." Coupling antennas on a tubular mast or a pylon.

- Chapter 6: "Cellular Networks: Antenna Tilt, Polarization Diversity, and Multiband on Panels." Use of the directional panels in cellular networks: "Tilt," Space diversity, Polarization diversity, multiband panels. Examples of industrial antennas.

- Chapter 7: "Filters." LC filters, microstrip line filters, coaxial cavity filters. Examples of industrial filters.

- Chapter 8: "Coupling of Several Transceivers on the Same Antenna." Coupling on one antenna of several transceivers. Examples of industrial couplers.
- Chapter 9: "Study of the Radio Environment on Antenna Site, Measurements of the Spurious Frequencies, and Protection Against These Spurious Frequencies."
- Chapter 10: "Radiating Apertures Horn and Parabolic Antennas." Calculation of the characteristics of radiating apertures UHF and SHF: application to pyramidal horns and to parabolic antennas for Microwave links. Example of an industrial parabolic antenna.

Part III: Appendixes
This part includes a number of useful reminders:

- Appendix A: Frequency bands in mobile radio networks
- Appendix B: Vector calculus
- Appendix C: Complex numbers
- Appendix D: Electrostatics
- Appendix E: Electromagnetism
- Appendix F: Physics of vibrations
- Appendix G: Transmission lines at high frequencies
- Appendix H: Waveguides

I wrote this book for engineers and technicians working in mobile radio neworks, and for students. I did my best in the study of antennas (Part I) to use the complex numbers when possible, which allow one to add electric fields by making sums of vectors.

I particularly thank Christian Harel, the managing director of Kathrein France, for contributing the Foreword and allowing me to present Kathrein antennas in this work. Thanks also go to his colleagues, Pascal Vallet, Henri Benhamou, Jean Fleury, and Dragan Gavric, for their technical input.

I also thank Christian Le Mazou for his advice on radioelectric measurements.

Conventions

The system unit used is the work is MKSA.

The function $y = A \sin(\omega t - \varphi)$, will very often be replaced by the complex function:

$$\bar{y} = A e^{j(\omega t - \varphi)} = A e^{j\omega t} e^{-j\varphi}$$

which makes it possible to separate the temporal term from the phase term (see complex numbers in Appendix C).

Part I:
Antennas in Mobile Radio Networks

1

Fundamentals of Antennas

1.1 Antenna History

We cannot begin a discussion on antennas without a reminder of the scientific discoveries made in the second half of the nineteenth century. These discoveries led to the first links of long-distance wireless telegraphy using wire antennas at the end of that century.

1.1.1 Maxwell Theory and Hertz Radiating System

Adding to the previous work on electricity and magnetism done by Michael Faraday and André-Marie Ampère, Scottish physicist and mathematician James Clerk Maxwell presented four differential equations bearing his name, which changed the existing knowledge on electrostatics and electromagnetism in 1864 at the Royal Society in Edinburgh, (see Appendixes D and E).

With these equations, Maxwell put forth that any source consisting of an alternating motion of electric charges in the air gives rise to electromagnetic waves propagated at the speed of light in all directions (the speed of light was already known at the time and was equal to 300,000 km/s). Each of these electromagnetic waves comprises an electric field and a magnetic field perpendicular to each other, and perpendicular to the direction of propagation (Figure 1.1).

Maxwell compares this phenomenon with that of a stone tossed into a pond. At the point of impact, the stone produces water oscillations, propagating on the surface of the water as waves (see Appendix F). Maxwell gives two examples of electromagnetic waves existing in space and explaining his

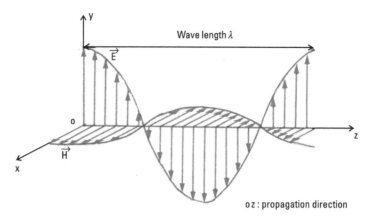

Figure 1.1   Electromagnetic wave.

theory: those caused by storms and those of light. From this discovery, many researchers have tried to imagine equipment able to generate electromagnetic waves. The first conclusive experiment is that of the German physicist Heinrich Hertz. In 1888, in his laboratory at the Technical High School in Karlsruhe (Germany), Hertz succeeded in generating electromagnetic waves in space with a *transmitter* (Figure 1.2), including a 6-volt battery and a Ruhmkorff coil, exciting an oscillating circuit that consisted of two copper balls (capacitor) and two copper rods (inductance) separated by a spark gap.

Due to a lack of sensitivity of the wave detector used, the *resonator*—which was a loop of copper wire with a spark gap—the propagation distance was limited to a few tens of meters. The operating mode is detailed below.

The high-frequency oscillator of several MHz (shown in Figure 1.3) consists of the capacitor C, which is formed by the two copper spheres A and

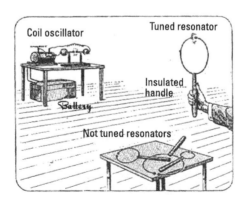

Figure 1.2   Charge characteristics for a battery.

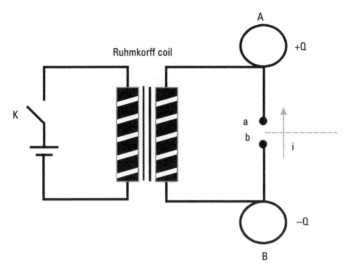

Figure 1.3 Hertz radiating system.

B (each with a diameter of 30 cm), and of the inductance of 3m copper rod connecting the two balls and cut in the middle by a spark gap composed of two small spheres a and b, the distance of which can be adjusted.

The large spheres are connected to the secondary coil of a Ruhmkorff coil, with the primary coil being connected to a 6-volt battery through a switch. The Ruhmkorff coil makes it possible to provide a voltage V of several kilovolts to the terminals of these two spheres at each closing of the switch. The switch is in fact the contact of a *trembler*, which is periodically opened and closed. At each closing of the contact of the trembler, the two spheres A and B become charged at $+Q$ and $-Q$, with $Q = V/C$.

From a certain value of V, sparks erupt between the small spheres a and b of the spark gap, causing an oscillatory discharge of the two spheres A and B, at a frequency of several MHz, through the oscillating circuit composed of the small inductance of the 3m copper rod and of the capacitor formed by the two large copper balls and whose dielectric is the atmosphere between these two balls. The current flowing in this oscillating circuit for the duration of the sparks comes in the form of a succession of damped sinusoids at a frequency of several MHz (Figure 1.4).

Moving the *resonator* in the axis of the spark gap of the oscillator and in several directions around the spark gap, Hertz observed sparkling at the spark gap of the loop, proving that his oscillator produced electromagnetic waves.

To produce sparkling on the spark gap of the resonator, it was indeed necessary that an electromotive force be created in the loop that could be

Figure 1.4   The current form in the oscillator.

achieved only if there was a variable magnetic field perpendicular to the plane of the loop (Lenz's law $e = -d\phi/dt$) or a variable electric field ($e = \int Edl$) integrated over the entire loop (see Appendixes D and E).

Hertz had concluded from the experiment that he had indeed verified Maxwell's theory, showing that electromagnetic waves could be created in the atmosphere by his spark gap, crossed by electric charges $q = \int idt$ oscillating as did the current in the oscillator (see Figure 1.5). That is not the only source; the electric charges in the dielectric (air) between the spheres A and B radiate as well.

1.1.2 Antenna Discovery

In order to increase the range of the Hertz radiating system, researchers have tried to improve the sensitivity of the wave detector.

The first result was obtained by French physicist, Edouard Branly. In 1890, Branly developed his *radio conductor*, which consisted of a closed cir-

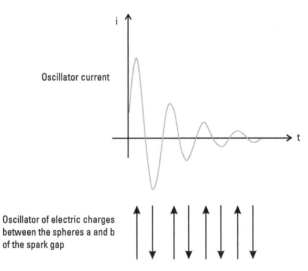

Figure 1.5   Oscillations of the electric charges in the spark gap.

cuit that contained a battery, a galvanometer, and a glass tube containing two pistons face-to-face with iron filings between them (Figure 1.6).

The reception of an electromagnetic wave decreases the electric resistance of the filings considerably, causing a current flow in the galvanometer.

The sensitivity obtained was 10 times higher than that of the *resonator*.

In 1895, Russian engineer Alexander Popov invented the first antenna to further improve the reception. For that purpose, he added to the radio conductor one inductance, two capacitors, and two access terminals. One terminal was connected to the ground, the other to a wire suspended from a wooden post. We will see further that the wire length was $\lambda/2$.

In 1898, French engineer Eugène Ducretet established the first wireless telegraph link in Paris, between the Eiffel Tower and the Pantheon (a distance of 4 km). The spark gap oscillator was connected to a wire antenna by the use of one inductance and two capacitors. We shall see later that this assembly was a matching circuit. The receiver was equipped with a radio conductor and a wire antenna.

This experiment showed that the transmitter antenna radiates.

In 1899, physicist Gugliemo Marconi was equipped with the same type of equipment as Ducretet, but adjusted at a lower frequency. After increasing the length of the receiver antenna and arranging both antennas vertically, the range increased significantly. Marconi was able to achieve a telegraph link of 45 km between Dover (UK) and Wimereux (France).

In 1901, Marconi made the first transatlantic telegraph link between Poldhu (Cornwall) and Newfoundland (Canada) by using a frequency between

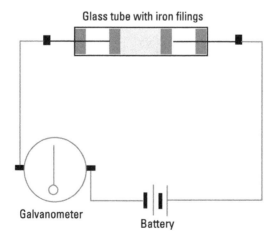

Figure 1.6   Radioconductor.

10 and 30 MHz to benefit from the reflections of the electromagnetic wave on the ionospheric layers. A vertical wire antenna was installed in Poldhu. This wire antenna was tuned to the frequency and a very powerful spark gap oscillator was used. It was able to provide several tens of kilovolts to the spheres A and B to supply high power to the antenna.

In Newfoundland, a radioconductor connected to a *kite antenna* was installed. This 122m-wire antenna was hung from a kite. The first antennas were thus born between 1895 and 1901, thanks to the work of these researchers (see Figure 1.9). From then on, antennas have been developed constantly in a number of models in order to meet a wide variety of needs.

1.2 How an Antenna Radiates

In the previous transmissions of Ducretet and Marconi, these researchers have demonstrated that a wire antenna radiates. This can be explained right now from the half-wave vertical *dipole*, which will be studied further and which is the basic element of antennas used in mobile radio networks.

This dipole consists of two aligned vertical metal strands, of length $\lambda/4$, connected to the transmitter by an access line, a matching circuit, and a coaxial cable (see Figure 1.7). This antenna is matched to the output impedance 50 Ohms of the transmitter by the matching circuit. The power of the transmitter is provided to the dipole antenna. There is no reflected power. The voltage standing wave ratio (VSWR) is equal to 1 (see Appendix G).

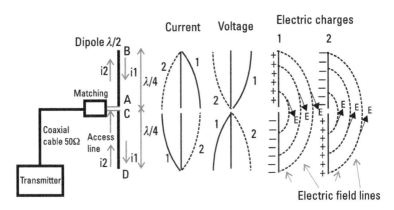

Legend: 1 = Quarter period number n of HF frequency at time t between i = o and i = i_{max}
        2 = Quarter period number n + 1 of HF frequencyc at time t between i = o and i = i_{max}

Figure 1.7   Half-wave vertical dipole.

The two $\lambda/4$ elements AB and CD are equivalent to an open quarter-wave line in which appear quarter standing waves of voltage and current (see Appendix F and G). We can see in Appendix G, in the case of an open line, the voltages on each wire of the line, at the same distance from the source, have a phase difference of 180°. It is the same for the currents.

According to this property, we have represented in the Figure 1.7, the voltage, the current, the electric charges in the antenna, for the quarter period n and the quarter period $n + 1$ of the frequency.

The electric charges create electric field lines going from positive charges to negative charges in the air in each half-plane around the antenna (see Appendix D).

Therefore, the direction of those electric field lines change at each quarter period of the frequency. It's the same for the *displacement currents* that follow the electric field lines (see Appendix D).

So, we have an oscillation of charges in the air around the antenna and, consequently, a production of electromagnetic waves.

Let us provide more details. Physicists have demonstrated in laboratory that the field electric lines form a closed loop tangential to the center of the dipole each time the current is $I_{max}(v = 0)$ in the dipole, that is to say once by half period of the frequency. Simultaneously the closed loop escapes the dipole in all directions. The electric field waves are therefore created by the electric fields of the closed loop (which are wavy) by the displacement currents around the antenna (see Section 1.3).

The electric waves are electromagnetic waves because Maxwell has demonstrated with his equations that if there is an electric field, there is always a magnetic field perpendicular to it (see Appendixes D and E).

Therefore, we can conclude that the electromagnetic waves are radiated from the center of the dipole.

1.3 Vertical Half-Wave Dipole Radiation Through Maxwell Equations

Figure 1.8 shows the dipole in a system of axis ox, oy, oz. We have chosen a point M on the axis oz to calculate the equation of the field wave E and of the field wave H in that direction.

The current i in the antenna creates in M a magnetic field $\vec{H}$, which, according to Laplace's law (see Appendix E), is parallel to ox and is represented at a given time in Figure 1.8. The current i in the antenna creates in M an electric field $\vec{E}$ is parallel to oy, because of the symmetry of the electric field lines

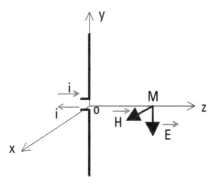

Figure 1.8   Representative drawing of the dipole.

(see above). It is represented on the figure at the same time. The components of the fields $\vec{E}$ and $\vec{H}$ on the three axes are thus:

$$\vec{E}\begin{vmatrix} 0 \\ -E_y \\ 0 \end{vmatrix} \qquad\qquad \vec{H}\begin{vmatrix} H_x \\ 0 \\ 0 \end{vmatrix}$$

The four Maxwell's equations are (see Appendixes D and E):

$$\vec{j} = \varepsilon\, d\vec{E}/dt$$

$$\vec{\nabla}\wedge\vec{E} = -d\vec{B}/dt$$

$$\vec{\nabla}\wedge\vec{H} = \vec{j} + \vec{i}$$

$$\vec{B} = \mu\vec{H}$$

where $\vec{j}$ is the displacement current per surface unit around the point M and $\vec{i}$ the conduction current. The last one is equal to zero because there is not any conduction current in the air.

So we can write $\vec{\nabla}\wedge\vec{H} = \vec{j}$

If we eliminate $\vec{B}$ and $\vec{j}$, we find:

$$\vec{\nabla}\wedge\vec{E} = \varepsilon\, d\vec{E}/dt \qquad\qquad (1.1)$$

$$\vec{\nabla}\wedge\vec{H} = \mu\, d\vec{H}/dt \qquad\qquad (1.2)$$

We can calculate the curls of $\vec{E}$ and $\vec{H}$ (see Appendix E):

$$\vec{\nabla} \wedge \vec{H} = \begin{vmatrix} \vec{i} & \vec{j} & \vec{k} \\ \dfrac{d}{dx} & \dfrac{d}{dy} & \dfrac{d}{dz} \\ Hx & 0 & 0 \end{vmatrix} \qquad\qquad \vec{\nabla} \wedge \vec{E} = \begin{vmatrix} \vec{i} & \vec{j} & \vec{k} \\ \dfrac{d}{dx} & \dfrac{d}{dy} & \dfrac{d}{dz} \\ 0 & Ey & 0 \end{vmatrix}$$

$$\vec{\nabla} \wedge \vec{H} = -d\frac{Hx}{dz}\vec{j} - \frac{dHx}{dy}\vec{k} = -\frac{dHx}{dz}\vec{j} \qquad \vec{\nabla} \wedge \vec{E} = -\frac{dEy}{dz}\vec{i} - \frac{dEy}{dx}\vec{k} = -\frac{dEy}{dz}\vec{i}$$

And we can write:

$$\frac{d\vec{E}}{dt} = \frac{dEx}{dt}\vec{i} - \frac{dEy}{dt}\vec{j} + \frac{dEz}{dt}\vec{k} = -\frac{dEy}{dt}\vec{j}$$

$$\frac{d\vec{H}}{dt} = \frac{dHx}{dt}\vec{i} + \frac{dHy}{dt}\vec{j} + \frac{dHz}{dt}\vec{k} = \frac{dHx}{dt}\vec{i}$$

On transferring these values in (1.1) and (1.2), we have

$$dHx/dz = \varepsilon\, dEy/dt \tag{1.3}$$

$$dEy/dz = \mu\, dHx/dt \tag{1.4}$$

By writing $d^2Ey/dzdt$ and $d^2Hx/dzdt$ in two ways, we find

$$d^2Ey/dz^2 - \varepsilon\mu\, d^2Ey/dt^2 = 0 \tag{1.5}$$

$$d^2Hx/dz^2 - \varepsilon\mu\, d^2Hx/dt^2 = 0 \tag{1.6}$$

These equations show that the E and H fields obey the second order differential equation of the waves (see Appendix F).

$$d^2U/dz^2 - 1/c^2\, d^2U/dt^2 = 0 \tag{1.7}$$

U is the wave amplitude and c is the speed of light.

We have thus found by calculation, taking the particular case of the direction oz, that:

1. The current in the antenna creates electric field and magnetic field.
2. The displacement current $\vec{j}$ around the point M gives rise to the electromagnetic wave. The displacement currents being distributed all around the antenna; the point M is necessarily closest the dipole.
3. The electromagnetic wave comprises two waves, the electric field wave and the magnetic field wave. In mobile radio networks, only the electric field wave is used.

So that confirms the previous explanations.

The area of the electric field lines represented in Figure 1.7, which extends over a distance of 1.5λ around the antenna (in the case of the half-wave dipole), is called the *Fresnel zone*. In this area, there are two types of electric and magnetic fields, those wavy by the displacement currents, each time $i = I_{max}$ and $v = 0$ (see Section 1.2), they are known as active, and the other, known as reactive.

From (1.3, (1.5), (1.6), and (1.7) we can deduce the following properties of the electromagnetic wave:

1.3.1 Electromagnetic Wave Velocity

If we compare (1.5) and (1.6) with (1.7), we obtain:

$$c = 1/\sqrt{\varepsilon\mu}$$

With the values of ε and μ of the MKSA system recalled in Appendixes D and E, we find $c = 300,000$ km/s.

1.3.2 Relationship between the Electric Field and the Magnetic Field

The solution of (1.7) is of the form

$$u = f(ct - z)$$

Assuming that we are interested only in the progressive wave (see Appendix F), which is always the case for the antennas.

The solutions of (1.5) and (1.6) would be thus of the form:

$$Hx = h(ct - z) \qquad Ey = e(ct - z)$$

Let's transfer these values in (1.3).

For that, let's calculate dHx/dz and dEy/dt by applying the general formula

$$df/dx = (df/du)(du/dx)$$

We find

$$dHx/dz = -dh/d(ct - z)$$
$$dEy/dt = cde/d(ct - z)$$

Therefore, we have $-dh/d(ct - z) = \varepsilon cde/d(ct - z)$ according to (1.3).
That is to say $h = -\varepsilon ce = -\sqrt{\varepsilon/\mu}\, e.$
Hence

$$Hx = -\sqrt{\varepsilon/\mu}\ Ey \qquad\qquad (1.8)$$

1.3.3 Electric and Magnetic Power of the Electromagnetic Wave: Poynting Vector

The power crossing a surface element dS of the space is:

$$dWe = \varepsilon(Ey^2/2)ds \qquad \text{for the electric power}$$
$$dWm = \mu(Hx^2/2)ds \qquad \text{for the magnetic power}$$

(see Appendixes D and E). It follows from (1.8) that at a given time a surface element dS is crossed by the same electric and magnetic power.

$$dWe = dWm$$

The total power passing through the surface element dS is:

$$dW = dWe + dW$$
$$dW = \varepsilon(Ey^2/2)dS + \mu(Hx^2/2)dS$$
$$dW = \varepsilon Ey^2 dS$$
$$dW = \sqrt{\varepsilon\mu}\, EyHxdS$$

The vector $P = \sqrt{\varepsilon\mu}\ \vec{E} \wedge \vec{H}$ is called *Poynting vector*, $\vec{E} \wedge \vec{H}$ being the vectorial product of $\vec{E}$ and $\vec{H}$ (see Appendix B).

Its flux through dS is $P \cdot dS = \sqrt{\varepsilon\mu}\ EyHxdS$.

The power dW, which crosses the surface element dS, is thus the flux of the Poynting vector through dS.

1.4 Wave Surface, Spherical Wave, Plane Wave, and Wave Polarization

The following sections provide several definitions concerning the electromagnetic waves.

1.4.1 Wave Surface

An antenna at a great distance can be considered as a point source. The wave surface is the locus of points in space reached at time t by the waves emitted at earlier time t_0 by a point source.

As $\Delta t = t - t_0 =$ constant for all the points of this surface, the latter is an equiphase surface (at all the points of the surface, the electric fields are in phase). As the propagation speed is the same in all directions, this wave surface is a sphere or a portion of a sphere.

1.4.2 Spherical Wave

As the wave surface of the waves emitted by a point source is a sphere or a portion of a sphere, then the point source is said to emit spherical waves.

An antenna on a pylon at a great distance, in mobile radio networks, is considered as a point source, therefore it emits spherical waves, with their wave surfaces being portions of sphere.

1.4.3 Plane Wave

A plane wave is a wave whose point source would be at infinite distance.

The wave surface would then be a *plane*. Far from an antenna and around a point on a short distance, the spherical wave surface can be confused with a plane. We then get a *plane wave*.

It should be noted that transmitting antennas in mobile radio networks emit spherical waves that are equiphase and nonequiamplitude, because they do not emit the same power in all directions (radiation patterns of the antennas). At a great distance from the transmitter, near the receiving antenna, the wave is still considered a plane.

The plane wave in this case remains *equiphase* and *equiamplitude*.

1.4.4 Wave Polarization

An electromagnetic wave has a vertical polarization when the electric field in the propagation direction is vertical.

An electromagnetic wave has a horizontal polarization when the electric field in the propagation direction is horizontal. The plane defined by the electric field and the propagation direction of the wave is called the *wave polarization plane*.

1.5 Electric Field Power Loss between an Antenna Transmitter and an Antenna Receiver in Free Space

We shall see in Chapter 2 that the electric field at distance r, emitted by a vertical half-wave dipole connected to the transmitter, is inversely proportional to r: $E = k/r$, k being a constant, E in Volt/meter, and r in meter.

Let's suppose that at a long distance r—in direct view from the transmitter—we have another vertical half-wave dipole connected to a receiver of input impedance 50 Ω. This antenna will receive from the transmitter antenna electric fields, which will provide an electromotive force to the receiver:

The electromotive force is

$$e = \int_{-\lambda/4}^{+\lambda/4} E dl \quad \text{(see Appendix C)}$$

Thus

$$e = E\lambda/2$$

E being constant and in phase in each point of the antenna (see Section 1.4.3)

The power at the receiver input is then in watts

$$P = e^2 / 200$$

Therefore,

$$P = k^2\lambda^2 / 800r^2$$

James Maxwell
1831–1879

Heinrich Hertz
1857–1894

Edouard Branly
1844–1940

Gugliemo Marconi
1874–1937

Alexandre Popov
1859–1906

Eugène Ducretet
1844–1915

Figure 1.9   Antenna researchers James Maxwell, Heinrich Hertz, Edouard Branly, Gugliemo Marconi, Alexandre Popov, and Eugène Ducretet.

The power received by the receiver is then inversely proportional to the square of the distance and to the square of the frequency. The propagation loss of the electric field, in direct view between an antenna transmitter and an antenna receiver, is therefore proportional to the square of the distance and to the square of the frequency.

1.6 Antenna Parameters

In order to define the antenna performances a certain number of parameters are required.

In practical terms, the main parameters are:

- The polarization;
- The radiation patterns;
- The gain;

- The aperture angle at half-power;
- The bandwidth;
- The input impedance.

1.6.1 Antenna Polarization

The polarization of an antenna is the polarization of the electromagnetic wave (see Section 1.4.4) emitted by the antenna in the direction of maximum radiation (case of the directional antenna) or in the directions of maximum radiation (case of the omnidirectional antenna). The antenna polarization will then be vertical if the electromagnetic wave polarization is vertical in that direction (or in those directions). The antenna polarization will then be horizontal if the electromagnetic wave polarization is horizontal in that direction (or in those directions).

1.6.2 Antenna Radiation Patterns

The antenna radiation patterns are made from the radiation intensity in every direction around the antenna.

1.6.2.1 Radiation Intensity in a Given Direction

We must start by making a reminder about the solid angle. For a circle in a plane (Figure 1.10), the radian is the central angle that subtends an arc of length equal to the radius r of the circle.

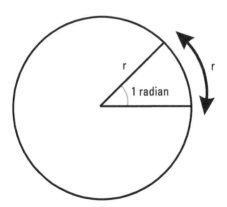

Figure 1.10 The radian.

If the subtended arc has a length L, the central angle is equal to L/r radians. If L is equal to the circumference $2\pi r$, the central angle is 2π radians. By analogy, let us consider a sphere of radius r (Figure 1.11).

The steradian is defined as the solid angle that subtends a spherical cap of surface equal to r^2.

If dS is a surface of the sphere, the solid angle that subtends it is equal to

$$d\Omega = dS/r^2 \text{ steradian} \tag{1.9}$$

For $S = 4\pi r^2$ (the surface of the sphere), the solid angle is 4π steradian. Let's assume that at the center of the sphere we have an antenna which is a radiating point source (see Section 1.4.1). The antenna radiation intensity in a given direction is defined as the power radiated by the antenna per unit of solid angle. It is expressed in *Watts per Steradian*.

Let W be the radiated power per square meter through a surface element dS surrounding a point M at distance r in the direction (θ, ϕ), in spherical coordinates (Figure 1.12).

The radiated power on the surface element dS will be WdS. It will also be the radiated power in the solid angle $d\Omega$ subtending dS. The radiation intensity in the direction (θ, ϕ) will then be: $U = WdS/d\Omega$. This concept is interesting because it does not involve the distance.

From (1.9), we can also write

$$U = Wr^2 \tag{1.10}$$

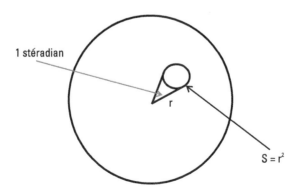

Figure 1.11 The steradian.

1.6.2.2 Application: Isotropic Antenna Radiation Intensity

An isotropic antenna is a fictitious point source that would radiate all the power provided by the transmitter in all directions. As a consequence, the radiated power per unit of solid angle (or radiation intensity) for an isotropic antenna is

$$U_0 = \frac{P_e}{4} \text{ expressed in watts/steradian} \qquad (1.11)$$

where P_e is the power provided to the antenna. However, is it interesting to have an antenna radiating in all directions (in all the solid angles of the sphere)? *No! And this will allow us to obtain gain.*

1.6.2.3 Radiation Patterns

The radiation pattern is a graphical representation of the properties of an antenna radiation. This graphical representation can be done in a system of coordinates with three axes ox, oy, and oz, using the spherical coordinates r, θ, ϕ (Figure 1.13). Let the antenna be in horizontal polarization and centered at the point O.

The measurements being made far from the antenna, let's replace the antenna by the point source O (see Section 1.4.1). In this case, the radiation pattern will be the representation in space of the antenna radiation intensity $U(\theta, \phi)$ in all directions (θ, ϕ).

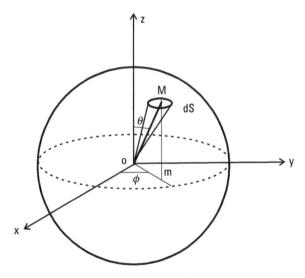

Figure 1.12 Surface element surrounding a point M in spherical coordinates.

Measurements are done on a small antenna proportional to the real antenna. They consist of using a field strength meter to determine the electric field E in a large number of points of a sphere of large radius r, and to calculate the power per square meter of this field by the formula $W = 1/2 \, \varepsilon \, E^2$ (see Appendix D).

Then we just have to calculate for each direction (θ, ϕ) the radiation by the formula $U = Wr^2$ [see (1.10)]. For practical reasons, the usual representation of the radiation pattern is made not in space but using two planes:

1. The E-plane.
2. The H-plane.

For *directive antenna* (see Figure 1.13), the E-plane is defined as the plane containing the direction of maximum radiation and the electric fields in that direction.

For *omnidirectionnal antenna*, the E-plane is defined as the plane containing any one direction of maximum radiation and the electric fields in that direction.

For *directive antenna*, the H-plane is defined as the plane containing the direction of maximum radiation and the magnetic fields in that direction.

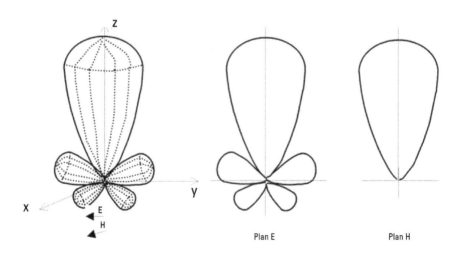

Figure 1.13   Radiation patterns.

For *omnidirectional antenna*, the H-plane is defined as the plane containing any one direction of maximum radiation and the magnetic fields in that direction.

When an antenna can be installed only in vertical polarization, the E-plane is often called (by manufacturers) the vertical plane, and the H-plane is often called the horizontal plane. Manufacturers use the representation E-plane and H-plane generally only when the antenna can be installed with either of the two polarizations. This avoids supplying four diagrams.

Example: Yagi Antenna

In the E-plane, the diagram obtained represents the variation of the radiation intensity $U = Wr^2$ for θ, varying from 0 to 2π.

In the H-plane, the diagram obtained represents the variation of the radiation intensity $U = Wr^2$ for ϕ, varying from 0 to 2π.

The radiation intensity being proportional to W, these representations indicate in dB ratios of radiated power relative to a maximum reference value (0 dB) for each point of a circle of radius r.

Instead of taking power ratios in dB, we can also take electric field ratios in dB. This is because these ratios are equal due to the relation $W = 1/2 \, \varepsilon \, E^2$. This is what we will do to calculate the radiation patterns in the next chapters.

The radiation pattern of an antenna is characterized by its radiation lobes:

- Main lobe;
- Secondary lobes;
- Rear lobe.

For the rear lobe of the directional antenna, the manufacturers provide the *front/back ratio* to indicate the level of attenuation of the radiation to the rear of the antenna. This attenuation is always greater than or equal to 25 dB.

1.6.3 Antenna Gain

1.6.3.1 Isotropic Gain

The isotropic gain (or absolute gain) of an antenna is the ratio of its maximum radiation intensity U_{max} to the radiation intensity, in the same direction, of an isotropic antenna that would radiate the input power of an antenna. If P_e is the power supplied to the antenna, the radiation intensity of the isotropic antenna is

$$U_i = P_e/4\pi$$

Therefore,

$$G_i = 4\pi U_{max}/P_e$$

For directional antennas, manufacturers usually provide the gain G_i. To calculate the power radiated in the favored direction, we simply apply the formula:

$$P_{rad} = G_i \times P_e$$

or

$$P_{rad} \text{ (dBW)} = G_i \text{ (dB)} + P_e \text{ (dBW) with W = Watts.}$$

1.6.3.2 Relative Gain

The relative gain is the gain compared to that of a reference antenna, generally the half-wave dipole. This gain is generally used by manufacturers for vertical omnidirectional antennas.

In this case, in order to calculate the radiated power, it is necessary to use the following formula:

$$G_i \text{ (dB)} = G_d \text{ (dB)} + 2.16$$

Indeed, we shall see in Chapter 2 that the isotropic gain of a half-wave dipole is:

$$G_i \text{ (dipole)} = 2.16 \text{ dB}$$

1.6.4 Aperture Angle

In the planes of radiation patterns E and H, the width of the half-power main lobe, or aperture angle of the antenna, is defined as the angle of two directions for which the radiation intensity is 3 dB lower than the maximum radiation intensity.

The smaller the aperture angle is, the larger the *antenna directivity* is.

1.6.5 Bandwidth

The bandwidth of an antenna is the bandwidth for which the VSWR (see Appendix G) is less than 1.5.

1.6.6 Antenna Input Impedance

Generally, the antenna input impedance is of the form:

$$Z_a = R_a + jX_a$$

  R_a is the active part

  X_a is the reactive part

R_a includes two components:

$$R_a = R_r + R_p$$

  R_r is the radiation resistance.

  R_p is the resistance of loss.

The antenna efficiency is defined as the following ratio:

$$\rho = (R_r/R_r + R_p)$$

As R_p is always very small compared to R_r, the efficiency of an antenna is at least about 95%.

   In order to have the maximum transfer of energy between the transmitter and the antenna or between the antenna and the receiver, the output impedance of the transmitter or the input impedance of the receiver antenna must be of the following form:

$$Z_t = R_a - jX_a \qquad Z_r = R_a - jX_a$$

The output impedance of the transmitter and the input impedance of the receiver being always 50 Ω, it is necessary to have a matching circuit, which transforms 50 Ω into $R_a - j\,X_a$, and vice versa. This circuit is integrated in the bottom of the antenna.

   Please note that the reactive part is important especially on the HF antennas (band 3–30 MHz). In that case, the matching circuit is complex. For the antennas we are interested in (very high frequency (VHF) (30–300 MHz) and ultrahigh frequency (UHF) (300–3000 MHz)], the reactive part is very low and the matching circuit is easy to do.

References

[1]   Balanis, C. A., *Antenna Theory, Second Edition*, Hoboken, NJ: Wiley, 1997.

[2]   Kraus, J. D., *Antennas for all Applications, Third Edition*, R. J. Marhefka (ed.), New York: McGraw Hill, 2002.

[3]   Stutzman, W. L., *Antenna Theory and Design, Second Edition*, G. A. Thiele (ed), Hoboken, NJ: Wiley, 1998.

[4]   Wolf, E. A., *Antenna Analysis*, Hoboken, NJ: Wiley, 1966.

2

Omnidirectional Vertical Wire Antennas

2.1 Introduction

The omnidirectional vertical wire antennas are mainly used in private mobile radio (PMR) networks, but they can also be used in Commercial Mobile Networks (2G, 3G, 4G) for cells with low traffic and large coverage in rural areas. In this chapter, we will only study the electric field waves provided by these antennas. This is because only the electric field of the electromagnetic wave is of interest in radio communications.

We shall begin by calculating the field provided by a small vertical wire element called an *infinitesimal dipole*, and then, in order to calculate the electric field of the wire antennas used, we will just need to integrate the resulting field on the whole length of the antenna.

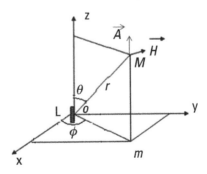

Figure 2.1   Infinitesimal dipole.

25

2.2 Electric Field of the Infinitesimal Dipole

In a system of rectangular axes ox, oy, and oz, in spherical coordinates, we have placed (see Figure 2.1) on the axis oz, a small conductor element L centered on o, and we will calculate the magnetic field at any point M of space produced by a high frequency current $i = I_0 \sin \omega t$, which we can also write as $i = I_0 e^{j\omega t}$ (see Appendix C). The conductor element is so short compared to the wavelength, that we can consider the amplitude I_0 constant along the wire.

We will calculate the potential vector $\vec{A}$ at the point M produced by the HF current i in the small wire, because we know that it is linked to the magnetic field by the relation

$$\vec{H} = \vec{\nabla} \wedge \vec{A} \qquad \text{(see Appendix E)}$$

and that the potential vector is of the form

$$\mathbf{A} = \vec{I}_0 \int_{-L/2}^{+L/2} \vec{d}_L / 4\pi r$$

We can then obtain the electric field with

$$\vec{H} = -\sqrt{\varepsilon/\mu}\,\vec{E}$$

Due to the propagation of the electromagnetic wave, the vector potential in M at time t is created by the current i at time $(t - r/c)$, where c is the velocity of the electromagnetic wave. We know that this potential vector is parallel to L. Thus, its components are

$$\vec{A} \left| \begin{array}{l} A_x = 0 \\[2ex] A_y = 0 \\[2ex] \overline{A}_z = \int_{-L/2}^{+L/2} I_0 e^{j\omega(t-\frac{r}{c})} d_L / 4\pi r = I_0 e^{j\omega(t-\frac{r}{c})} L / 4\pi r \end{array} \right.$$

$\overline{A}_z$ is a complex number (see Appendix C).

The curl of $\vec{A}$ is equal to

$$\vec{\nabla} \wedge \vec{A} = \begin{vmatrix} \vec{i} & \vec{j} & \vec{k} \\ \dfrac{d}{dx} & \dfrac{d}{dy} & \dfrac{d}{dz} \\ 0 & 0 & A_z \end{vmatrix}$$

Thus, the components of the field $\vec{H}$ are

$$\vec{H} \begin{vmatrix} H_x = dA_z/dy \\ H_y = dA_z/dx \\ H_z = 0 \end{vmatrix}$$

Taking the derivatives of $\overline{A}_z$ with respect to x and to y, given by $r = \sqrt{x^2 + y^2}$ we find terms in $1/r$, $1/r^2$, $1/r^3$. At a great distance from the antenna, the last two terms are negligible. We therefore only keep the terms in $1/r$.

Thus, we finally find

$$\vec{H} \begin{vmatrix} \overline{H}_x = -j\dfrac{\omega}{c}\dfrac{y}{r} I_0 \, L/4\pi r e^{j\omega(t-\frac{r}{c})} \\\\ \overline{H}_y = j\dfrac{\omega}{c}\dfrac{x}{r} I_0 \, L/4\pi r e^{j\omega(t-\frac{r}{c})} \\\\ H_z = 0 \end{vmatrix}$$

This field is perpendicular to OM and parallel to the plane xoy. Indeed, the field H is parallel to the plane xoy because $H_z = 0$, and H is perpendicular to OM because the scalar product of vector $\vec{H}$ and vector $\overrightarrow{OM}$ is equal to zero (calculus of scalar product with the coordinates of $\vec{H}$ and $\overrightarrow{OM}$—see Appendix B).

Its amplitude is

$$H = \sqrt{Hx^2 + Hy^2}$$

$$H = \dfrac{\omega}{c} I_0 L \sqrt{x^2 + y^2}/4\pi r^2$$

As $\sqrt{x^2 + y^2}/r = \text{Om}/\text{OM} = \sin\theta$ and $\omega/c = 2\pi/\lambda$ (see Appendix F), we can write

$$H = I_0 L \sin\theta/2\lambda r \qquad A/m$$

We have seen that the electric field was linked to the magnetic field by the relation

$$H = \sqrt{\varepsilon/\mu}E$$

which can be written as $H = \varepsilon\, c\, E$. Therefore, the amplitude of the electric field in M is

$$E = I_{0L}\,\sin\theta/2\varepsilon c\lambda r \qquad V/m$$

Let us replace ε and c by their values (see Appendix D and Chapter 1)

$$E = 60\pi I_{0L}\,\sin\theta/\lambda r \qquad V/m$$

The propagation equation of the field E in the direction OM is written in the vertical plane zoM in polar coordinates

$$\bar{E}(\theta,r,t) = 60\pi I_0\, L\,\sin\theta e^{j\omega(t-\frac{r}{c})}/\lambda r$$
$$\bar{E}(\theta,r,t) = 60\pi I_0\, L\,\sin\theta e^{j(\omega t-2\pi r/\lambda)}/\lambda r \qquad (2.1)$$

after having replaced $\omega\, r/c$ by $2\pi\, r/\lambda$.

This equation is the basic equation for wire antennas.

It may be noted that we find the phase term $2\pi r/\lambda$, which indicates, in radians, the phase delay of the electromagnetic wave at the distance r (see Appendix F). To draw the radiation pattern in the vertical plane (plane oz, OM), we have seen in Chapter 1 that we could replace the radiation intensity $U(\theta)$ by the electric field $E(\theta)$, provided that we replace the power ratios in decibels by electric field ratios in decibels. According to (2.1), the only term that depends only on θ is sin θ. The representative curve in the vertical E-plane plane of $E(\theta)$ therefore has as an expression the following *characteristic function*

$$F(\theta) = \sin\theta$$

This diagram is represented by two symmetrical circles relative to oz, as shown in Figure 2.2.

In the horizontal plane, the radiation pattern is obviously a circle. This is because there will always be the same diagram in the vertical plane if Φ is varied from 0 to 2π (Figure 2.3).

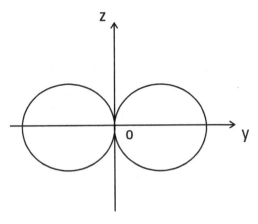

Figure 2.2   Vertical plane (E-plane).

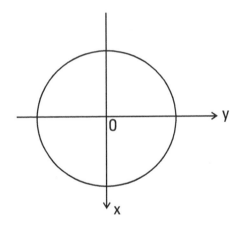

Figure 2.3   Horizontal plane (H-plane).

2.3 Image Theory—Ground Influence

Let a vertical dipole be at a distance d above a perfectly conducting plane surface (see Figure 2.4).

A point M of space will receive two waves: a direct wave and a wave reflected by the surface. The theory of images indicates that this scheme is equivalent to two dipoles in space to the distance $2d$ in the vertical plane and travelled by currents in phase opposition. Therefore the reflection coefficient of the surface is equal to 1. The power at the antenna input is divided by half on the dipole and its image.

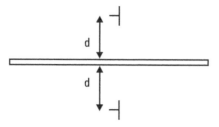

Figure 2.4   Conducting plane surface influence.

If the conducting plane surface is replaced by the ground the reflection coefficient is less than one. In practice the influence of ground is real only if the distance between the antenna and the ground is less than 10λ. The influence of ground does not modify the radiation pattern in the horizontal plane.

In the vertical plane, the presence of ground reduces the aperture angle of the antenna, which can be annoying in some cases: mountain area, urban area. Therefore, it is recommended to always install the antennas from the ground at a distance greater than 10λ.

2.4   Omnidirectional Vertical Wire Antennas

A wire antenna is any straight antenna, either horizontal or vertical.

We will limit ourselves to the study of vertical antennas that are only used in mobile radio because we have to use vertical polarization. We will only calculate the electric field that is also the only one used in mobile radio. The antennas will always be isolated from the ground, that is to say, at more than 10λ from the ground, and we shall assume them without loss.

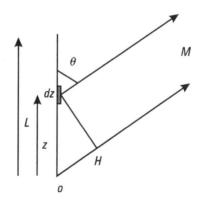

Figure 2.5   Straight vertical antenna isolated in space.

Let us suppose a radiating vertical wire of length L is isolated from the ground (Figure 2.5). The wire is supposed to be powered in the bottom at the point O by a current $i = I_0 e^{j\omega t}$.

The total field radiated at a remote point M, in the direction θ, is the sum of the fields radiated by each element dz.

According to the theory of lines (see Appendix G), the current in the element dz at the distance z from O has the following value depending of the maximum current amplitude I_M of the standing wave at this time:

$$i_z = I_M \sin\beta(L - z)$$

where $\beta = 2\pi/\lambda$.

The electric field radiated in M by the element dz located in O is

$$d\bar{E}_z = 60\pi I_0 dz \sin\theta e^{j(\omega t - 2\pi r/\lambda)}/\lambda r \qquad \text{[please see (2.1)]}$$

with r being the distance between dz and M. The electric field dE_z radiated in M by the current i_z of each element dz, located at the distance z from O, will have, relative to the field dE_0, a phase lead of

$$2\pi OH/\lambda = 2\pi z \cos\theta/\lambda = \beta z \cos\theta$$

corresponding to the course difference of the two waves. We obtain

$$d\bar{E}_z = 60\pi I_M \sin\beta(l - z)dz \sin\theta e^{j\beta z \cos\theta} e^{j(\omega t - 2\pi r/\lambda)}/\lambda r$$

To obtain the electric field, dE_z must be integrated over the entire length L of the antenna wire.

$$\bar{E} = 60\pi I_M \sin\theta e^{j(\omega t - 2\pi r/\lambda)}/\lambda r \int_0^L \sin\beta(L - z)e^{j\beta z \cos\theta} \, dz$$

The calculation of this integral is long and of no interest. We find

$$\bar{E} = (60 I_M/r\sin\theta)\{\cos[(\beta_L \cos\theta)/2] - \cos\beta L/2\}e^{j(\omega t - 2\pi r/\lambda)} \qquad (2.2)$$

We will calculate the value of the electric field and draw the radiation patterns in the vertical plane (E-plane) for $L = \lambda/2$, $L = \lambda$, $L = 3/2\,\lambda$.

2.4.1 Half-Wave Antenna

The modulus value of the electric field is

$$E = 60I_M \cos(\pi/2 \, \cos\theta)/r\sin\theta \tag{2.3}$$

The electric field equation is

$$\bar{E} = 60I_M \cos(\pi/2 \, \cos\theta)/r\sin\theta \, e^{j(\omega t - 2\pi r/\lambda)} \tag{2.4}$$

The *characteristic function* is:

$$F(\theta) = \cos(\pi/2 \, \cos\theta)/\sin\theta \tag{2.5}$$

Table 2.1 allows for the plotting of the radiation pattern in the vertical plane.
We have varied only θ between 0 and 90°, because if we change θ to $-\theta$ in (2.5), we would obtain the symmetrical curve relative to the horizontal axis. If we rotate the E-plane of 180°, we would obtain the symmetrical curve relative to the vertical axis.

It will thus be enough, after the plotting of the curve between $\pi/2$ and 0, to take the symmetrical curve relative to the horizontal axis, then the symmetrical curve of the whole set relative to the vertical axis.

The aperture angle corresponding to $F(\theta) = 0.7$ is equal to $2(90 - 51) = 78°$.

The radiation pattern in the vertical plane (E-plane) is represented in Figure 2.6.

The pattern in the horizontal plane (H-plane) would be a circle because, if we rotate the vertical plane formed by the wire and the point M around the

Table 2.1
$F(\theta)$ for $l = \lambda/2$

$\theta°$	$F(\theta)$
90	1
75	0.95
60	0.8
51	0.7
45	0.63
30	0.42
15	0.21
0	0

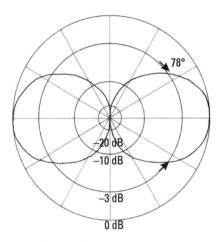

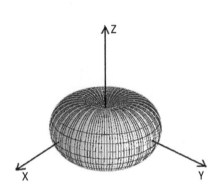

Figure 2.6 Radiation pattern of the half-wave antenna—E-plane.

Figure 2.7 Half-wave antenna radiation pattern in space.

wire, the radiation pattern shown above does not change. A three-dimensional representation of the half-wave antenna radiation pattern is given in Figure 2.7.

2.4.2 λ-Length Antenna (Full Wave)

The modulus value of the electric field is

$$E = 60I_M[\cos(\pi\cos\theta) + 1]/r\sin\theta$$

The characteristic function is

$$F(\theta) = [\cos(\pi\cos\theta) + 1]/\sin\theta$$

Table 2.2 allows for the plotting of the radiation pattern in the vertical plane. As previously stated, we have varied only θ between 0 and 90° for the same reasons. The aperture angle corresponding to $F(\theta) = 0.7$ is equal to 2 $(90 - 66) = 48°$. The radiation pattern in the vertical plane is represented in Figure 2.8.

For the same reason as for the half-wave antenna, the pattern in the horizontal plane (H-plane) would be a circle.

A three-dimensional representation of the radiation pattern of the antenna of length λ is given in Figure 2.9.

Table 2.2
$F(\theta)$ for $l = \lambda$

$\theta°$	$F(\theta)$
90	2
75	1.73
66	1.41
60	1.15
51	0.7
45	0.56
30	0.17
15	0.02
0	0

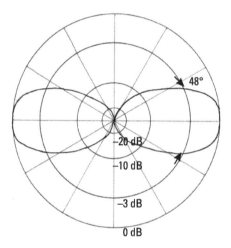

Figure 2.8 Radiation pattern of the λ length antenna (E-plane).

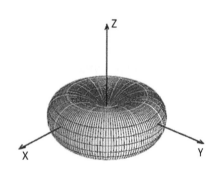

Figure 2.9 Radiation pattern of the λ length antenna in space.

2.4.3 1.5 λ-Length Antenna (Three Half-Waves)

The modulus value of the electric field is

$$E = 60I_M \cos(3\pi/2 \; \cos\theta)/r\sin\theta$$

The characteristic function is

$$F(\theta) = \cos(3\pi/2 \; \cos\theta)/\sin\theta$$

Table 2.3
F(θ)

θ°	F(θ)	θ°	F(θ)
90	1	40	1.39
85	0.92	35	1.31
80	0.68	30	1.18
75	0.33	25	1.01
70	0.04	20	0.82
65	0.45	15	0.61
60	0.81	10	0.41
55	1.1	5	0.21
50	1.3	0	0
45	1.39		

Table 2.3 allows for the plotting of the radiation pattern in the vertical plane. As previously, we have only varied θ from 0° to 90° for the same reasons. The radiation pattern in the vertical plane is represented in Figure 2.10.

For the same reasons as for the half-wave antenna, the pattern in the horizontal plane (H-plane) would be a circle. A three-dimensional representation of the radiation pattern of a 1.5 wave antenna is given in Figure 2.11.

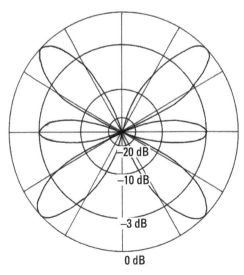

Figure 2.10   1.5λ antenna radiation pattern (E-plane).

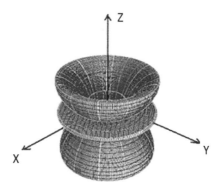

Figure 2.11   1.5λ antenna radiation pattern in space.

2.4.4 Result Analysis

The first finding is the presence in the vertical plane, for the 3λ/2 antenna, of two main lobes in a direction parallel to the ground and of four secondary lobes that share about 90% of the total power transmitted. As the only lobes of interest for us in mobile radio are the lobes in a direction that is parallel to the ground, this antenna is not interesting. It will be the same for all wire antennas of length greater than λ, with the number of their secondary lobes increasing with the length.

The second observation is that the antennas of half-wave and full-wave type have no secondary lobes in the vertical plane, but only two main lobes in a direction that is parallel to the ground. Therefore, they would be perfectly adapted for use on mobile radio infrastructure. However, the input impedance of the antenna of length λ has a real part of 2000 Ω, while the antenna of length λ/2 has an input impedance whose real part is 73 Ω, with the condition that the half-wave antenna is powered in the center (see Section 2.5).

There would be high voltages on the antenna of length λ, which would complicate its installation, not to mention the difficulties of matching to 50 Ω. This is why this antenna is never used. The only vertical wire antenna used in mobile radio is therefore the dipole, which is used all alone or in array antenna of dipoles for greater gains. A small exception, however, is the *ground plane* vertical antenna represented in Figure 2.12.

This antenna consists of a radiating strand of length λ/4, perpendicular to a ground plane consisting of four conductive strands of length λ/4, called radians, forming four right angles between them. According to the image theory, it is as if there was a λ/4 strand symmetrical to the radiating strand relative to the ground plane and travelled by a current in phase opposition. The dipole is thus reconstituted.

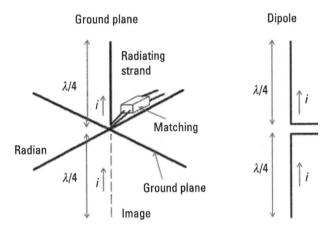

Figure 2.12 Ground plane antenna.

The ground plane antenna therefore has the same characteristics (gain and radiation patterns) as the dipole. For space reasons, the radians are usually inclined downwards, which does not change the antenna characteristics. This antenna is only used in private networks (PMR) in the VHF bands (see Appendix A).

2.5 Input Impedance of Vertical Wire Antennas

Let A and B be the two input terminals of a vertical wire antenna (Figure 2.13); one of the terminal may be grounded.

If v is the voltage between the two terminals A and B, and i the input current, the input impedance of the antenna is defined by the ratio

$$\bar{Z}e = v/i = R_A + jX_A$$

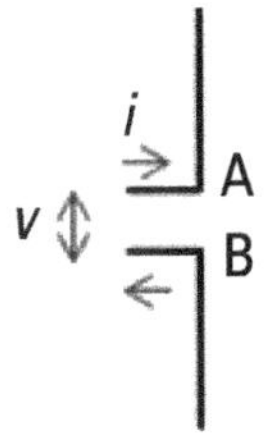

Figure 2.13 Wire antenna.

As we have seen in Chapter 1, the transmitter or the receiver must have a conjugate impedance $R_A - jX_A$ of the antenna impedance so that there is optimum transfer of power between the two devices. As in mobile radio, the output impedance of a transmitter and the input impedance of a receiver are always equal to 50 Ohms, this optimization is performed by a matching circuit.

Input impedance measurements were made experimentally on a wire line of a given length L, with ground base (B-grounded) by varying the frequency. The measurement result is represented in the complex plane Figure 2.14.

When L is small compared to the wavelength (case of high frequency antennas in the band 2–30 MHz), the input impedance is capacitive. The impedance becomes real at the P point for a value of L close to $\lambda/4$ ($X_A = 0$).

The antenna is said tuned and the wavelength is called *resonance wavelength*.

For $L = \lambda/4$, we find

$$Z_e = 36.6 \text{ Ohms} + j21.25 \text{ Ohms}$$

The impedance is inductive. This is the quarter-wave antenna shown in Figure 2.15 without its matching circuit to 50 Ω, and on which we have represented the standing quarter wave of current, with its antinode ($i = I_M$) at the foot of the antenna and its node ($i = 0$) at the other end. This antenna is used mostly on vehicles in mobile radio.

It is used on PMR infrastructure in the VHF band in the form of the *ground plane* antenna. If we continue to reduce the wavelength to the point Q close to $l = \lambda/2$, the impedance becomes real and very large (several thousand Ohms). The antenna is said tuned to the antiresonance.

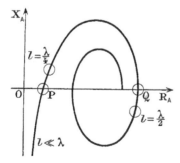

Figure 2.14 Input impedance of a vertical wire antennas of length L, with ground base, depending on the frequency.

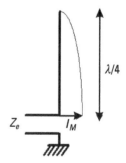

Figure 2.15 Quarter-wave antenna with ground base.

For $l = \lambda/2$, the impedance is still very large but capacitive.

Figure 2.16 represents this antenna with the current in the antenna. The matching circuit is not represented. It should be noted that the voltage at the foot of the antenna is reduced because the capacitive reactance of the input impedance is cancelled by the matching circuit.

Another way to power this antenna is to attack the center (current anti-node) because the impedance is lower (Figure 2.17), and easier to match. Thus we obtain the dipole with the electric field equation (2.4).

We have represented in Figure 2.17 only the current amplitude and not its phase. The measurement result in these conditions as

$$Z_e = 73.2 \text{ Ohms} + 42.5 \text{ Ohms}$$

The real part and the reactive part of the impedance of this antenna are twice those of the quarter-wave antenna. This is easily understandable because the antenna is equivalent, for its input, to two quarter-wave antennas in series. This antenna is easily adapted to 50 Ω by a circuit whose technology depends on the frequency, discrete elements (inductor and capacitor), and microstrip lines (see Appendix G).

That is why the half-wave antenna is most often used in this form.

Generally, the two central points of the antenna connected to the match-ing circuit are very close (a few millimeters), which is safe because the voltage at these points is zero (the phase shift between the current and the voltage is $\pi/2$). This has the effect of bringing closer the connecting wires, preventing them from radiating (see Appendix G or Chapter 8, Section 8.2).

Finally, impedance measurements were made on an antenna of length λ connected at the center, like the dipole, and the result is a real part of 2000 Ω.

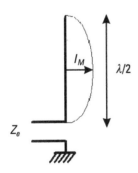

Figure 2.16 Half-wave antenna powered at the foot.

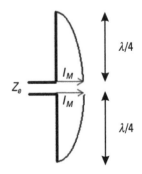

Figure 2.17 Half-wave antenna powered at the center.

2.5.1 Resonance and Antiresonance

It may be surprising that the resonance and antiresonance (the reactive part of the impedance equal zero) do not occur, respectively, for $l = \lambda/4$ and $l = \lambda/2$. This is because the propagation speed of the current and voltage waves in a conductor is slightly less than the speed of the electromagnetic wave in the air (290,000 Km/sec in the copper). This is also true for the wavelength. The result is that there are two ways to resonate a wire at $\lambda/4$ or $\lambda/2$:

- By calculating the length of the wire with the wavelength in the wire. In this case, there is a matching to be made between a pure resistance (the antenna) and 50 Ω (output transmitter or input receiver).

- By calculating the length of the wire with the wavelength in the air. In this case, there is a matching to be made between an impedance $R + jX$ and 50 Ω.

2.6 Gain Calculation of a Vertical Half-Wave Dipole Antenna Isolated in Space

2.6.1 Isotropic Gain G_i

Assuming that the vertical half-wave dipole is without loss, the total power P_r radiated by the antenna is equal to the power P_e provided to the antenna. Therefore, this power P_r is also the power radiated by an isotropic antenna that would radiate the power P_e. We have seen in Chapter 1 that the isotropic gain of an antenna in the direction of maximum radiation intensity is the ratio of this intensity to the radiation intensity in the same direction of an isotropic antenna that would radiate the power P_e.

As $P_e = P_r$, we can write

$$G_i = 4\pi U_{max}/P_r$$

Let's take a point M in a system of spherical coordinates, represented in Figure 2.18, and a dipole placed at the point O, in vertical polarization (directed according to oz).

The electric field created by the dipole at the remote point M at distance r in the direction Θ is given by (2.3):

$$E = 60I_M \cos(\pi/2\cos\theta)/r\sin\theta$$

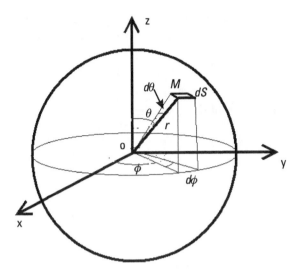

Figure 2.18 Point *M* in spherical coordinates.

The radiated power at this point per unit surface is $W = 1/2\ \varepsilon E^2$ (see Appendix D).

$$W = 1/2\varepsilon E^2 = 1/2\varepsilon 60^2 I_M^2 \cos^2(\pi/2\cos\theta)/r^2 \sin^2\theta$$

The total power radiated by the dipole in all directions is obtained by integrating W on the sphere of radius r. Let dS be a surface element around M. The elementary surface can be written as

$$dS = r^2 \sin\theta\, d\theta d\Phi$$

The arcs surrounding dS being respectively $r\, d\theta$ and $r\sin\theta\, d\Phi$. The power radiated in the dS element is

$$dP_r = 1/2\varepsilon 60^2 I_M^2 \left[\cos^2(\pi/2\cos\theta)\right]/\sin\theta\, d\theta d\Phi$$

By integrating dP_r over the entire surface of the sphere, we find

$$P_r = \pi/2\varepsilon 60^2 I_M^2 \times 2,435$$

The radiation intensity of the dipole in the direction $\theta = \pi/2$ (direction of maximum radiation intensity and that we are interested in mobile radio) is

$$U_{max} = Wr^2 = 1/2\varepsilon\,60^2\,I_M^2$$

The isotropic gain of the dipole is

$$G_i = 4\pi U_{max}/P_r = 4/2.435 = 1.643$$

Hence

$$G_i = 10\log 1.643 = 2.16\,dB \qquad (2.6)$$

2.6.2 Gain Relative to the Dipole

The gain relative to the dipole is obviously

$$G_d = 0 \text{ dB}$$

2.7 Half-Wave Antenna Alignment in the Vertical Plane

2.7.1 Principle

To obtain high gain antennas, several dipoles are grouped on a vertical axis. Two technologies are used:

1. Installation of vertical dipoles aligned along a pylon.
2. Integration of dipoles in a vertical fiberglass tube.

2.7.2 Vertical Dipoles Along a Pylon

This is the oldest technology. Groups of two or four dipoles interconnected by coaxial cables 50 Ohms can still be found on the mobile phone sites of PMR networks. The four-dipole antenna is represented in Figure 2.19.

The transmitter is connected with a coaxial cable 50 Ω to the coupler input. The four dipoles are vertically aligned and at a distance λ, a necessary condition to have a negligible coupling between antennas, as we shall see in Part II of this book.

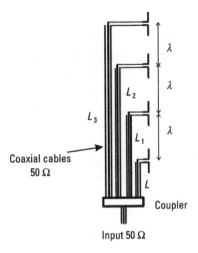

Figure 2.19 Four dipoles aligned on a pylon.

The four dipoles are connected by four coaxial cables coming from a 1-input/4-outputs coupler (equivalent to the 4-inputs/1-output coupler—see Part II). The different lengths of coaxial cables are linked by the following relationships:

$$L_1 = L + \lambda$$
$$L_2 = L + 2\lambda$$
$$L_3 = L + 3\lambda$$

Thus the four dipoles are out of phase with each other of 2π. We can consider that they are connected in phase, because that is not inconvenient for the modulation.

Each dipole receives a quarter of the transmitter power (−6 dB), which requires a matching circuit at the antenna input (coupler). The matching of each dipole to its coaxial cable is not represented.

2.7.3 Integration of Dipoles in a Fiberglass Tube

This technology, the most widely used at present, makes high-gain industrial antennas easier to install.

In this case, the dipoles are generally composed of copper tubes of length $\lambda/4$, filled with an insulating material with a wire through the centre of the tube. The tubes are equivalent to a coaxial cable with a characteristic impedance of 50 Ω (see Appendix G). We obtain thus an antenna commonly called *candle antenna*.

2.7.3.1 Example: Four Dipoles

The principle of connection is represented in Figure 2.20.

The dipoles consist of two elements $\lambda/4$, aligned side by side, but isolated, and connected, as represented on the left side of Figure 2.20.

The phase shift of λ between each dipole is carried out physically thanks to a spacing of $\lambda/2$ between the ends of the two adjacent dipoles, which allows them to have a negligible coupling between the dipoles (see Part II of this book).

The antenna behaves like an open line with current standing waves, which we have detailed at time t on two dipoles and the connection between these two dipoles (the right side of Figure 3.20).

This connection with its apparent two wires has a half-wave current on each wire, which are out of phase of 180° (see Appendix G or Chapter 8, Section 8.2). As these two wires are very close ($d \ll \lambda$), the electric fields created by these currents cancel each other out.

Only the currents in the tubes of the dipoles make the antenna radiate because the inner wires are unable to radiate. As the gap between the centers of the dipoles is equal to λ, the dipoles radiate with an outphase of 2π as previously. A matching of the antenna is required at the input. The four dipoles share the power of the transmitter.

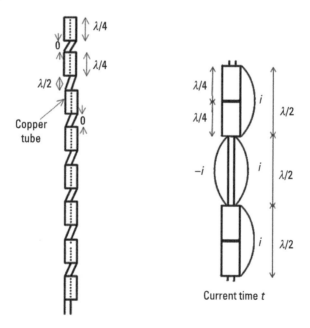

Figure 2.20   Integration of dipoles in a fiberglass tube.

It should be noted that the half-wave resonance of the dipoles actually consists of two very close quarter-waves, as we have seen so far. However, the gap is very small, and therefore invisible in the drawing. This is because of the very high ratio between the quarter-wave length and the thickness of the insulator.

The same principle of integration allows to provide vertical antennas of 2–6 dipoles, giving isotropic gains from 5–9 dB. Antennas with more gain are not manufactured because the aperture angle in the E-plane is too small (see Appendix 2A).

We have thereby all a series of *candle antennas* in the bands 150 and 450 MHz very much used in private mobile radio (PMR), and in the bands 900 and 1800 MHz for GSM.

2.7.4 Gain and Radiation Patterns of the Four-Dipole Vertical Antenna

We will calculate the gain G_d relative to the dipole and the isotropic gain G_i, and we will represent the radiation pattern in a vertical plane (E-plane). Figure 2.21 shows that the four dipoles are separated by λ in the E-plane, and a remote point M in this plane.

As we have explained in Chapter 1, the dipoles are radiating from the center. The electric field provided by the dipole A_1 at a point M faraway at the distance r in the direction θ is [see (3.4)].

$$\bar{E}_1 = 60 I_M \cos(\pi/2 \cos\theta)/r \sin\theta \, e^{j(\omega t - 2\pi r/\lambda)}$$

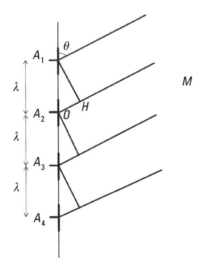

Figure 2.21 Four dipoles vertically aligned—vertical plane (E-plane).

The electric field supplied by the dipole A_2 at the point M is late relative to E_1 of $2\pi\, OH/\lambda$.

Using complex numbers, we can write

$$\overline{E}_2 = E_1 e^{-j(2\pi OH/\lambda)} = E_1 e^{-j(2\pi\cos\theta)}$$

because $OH = \lambda\cos\theta$.

The total field supplied by A_1 and A_2 is

$$\overline{E}_{2d1} = \overline{E}_1 + \overline{E}_2$$

This field can be represented in the complex plane, Figure 2.22, by the geometric sum of two vectors $\vec{E}_1$ and $\vec{E}_2$. The arrow indicates the direction of the phase lead.

Hence the modulus of $\vec{E}_{2d1}$

$$\mathbf{E}_{2d1} = 2E_1 \cos(\pi\cos\theta)$$

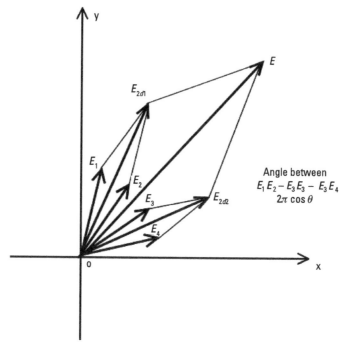

Figure 2.22 Complex plane.

We would find the same result for the dipoles A_3 and A_4

$$\mathbf{E}_{2d2} = 2E_1 \cos(\pi \cos\theta)$$

The total field $\vec{E}$ provided in M by the four dipoles is the geometric sum of two vectors $\vec{E}_{2d1}$ and $\vec{E}_{2d2}$. These two vectors form between them an angle equal to $4\pi \cos\theta$. Hence, the modulus of the total field

$$E = 2E_{2d1} \cos(2\pi \cos\theta)$$

By replacing $\vec{E}_{2d1}$ by its value

$$E = 4E_1 \cos(\pi \cos\theta)\cos(2\pi \cos\theta) \tag{2.7}$$

We call network factor in the vertical plane (E-plane) the quantity

$$F = 4\cos(\pi \cos\theta)\cos(2\pi \cos\theta) \tag{2.8}$$

2.7.4.1 Gain Relative to a Dipole and Isotropic Gain

Setting $\theta = \pi/2$ in (2.7), we find the maximum electric field $E = 4E_1$. Therefore the direction $\theta = \pi/2$ is the direction of the maximum radiation intensity $U_{\max}$. So let's calculate now the isotropic gain G_i.

The gain in electric field in the direction $\theta = \pi/2$ is 4. Therefore, the power gain relative to the dipole is 16 (12 dB). This is because the power is proportional to the square of the electric field ($W = 1/2\,\varepsilon\,E^2$). As the power input of each dipole is the quarter of the power at the foot of the antenna

$$G_d = 6 \text{ dB}$$

Therefore

$$G_i = 6 + 2.16 = 8.16 \text{ dB}$$

2.7.4.2 Radiation Patterns

The value of the total field at the point M can be written, replacing E_1 by its value in (2.4)

$$\overline{E} = 240 I_M \cos(\pi/2\cos\theta)\cos(\pi \cos\theta)\cos(2\pi \cos\theta)/r\sin\theta\, e^{j(\omega t - 2\pi r/\lambda)}$$

Table 2.4
The Characteristic Function (E-Plane)

$\theta°$	$F(\theta)$	$\theta°$	$F(\theta)$
90	1	40	0.04
85	0.81	35	0.17
80	0.38	30	0.24
75	0.04	25	0.28
70	0.24	20	0.24
65	0.18	15	0.19
60	0	10	0.11
55	0.16	5	0.07
50	0.19	0	0
45	0.1		

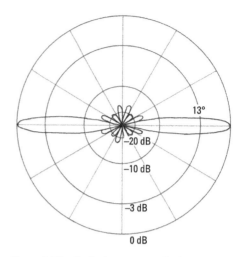

Figure 2.23   Radiation pattern of a four-dipole antenna (E-plane).

The characteristic function enabling to draw the radiation pattern in the E-plane is

$$F(\theta) = \cos(\pi/2\cos\theta)\cos(\pi\cos\theta)\cos(2\pi\cos\theta)/\sin\theta$$

You can therefore establish Table 2.4 allowing to plot the radiation pattern in the E-plane.

As previously, we have varied only θ from 0° to 90° for the same reasons.

The radiation pattern in the horizontal plane (H-plane) would obviously be a circle. The network factor in the H-plane is obtained by setting $\theta = \pi/2$ in (2.7). We find $F = 4$. It is interesting to compare the radiation pattern in the E plane with that of the dipole alone (see Figure 2.6).

We thus find that the gain results in a reduction of the aperture angle (in our example from 78° to 12°). This confirms the conclusion of Section F.5 of Appendix F, "Radiation of point sources aligned, with the same spacing, and in phase."

References

[1]   Badoual, R., *Microwaves, Volume 2*, S. Jacquet (ed), France: Masson, 1995.

[2]   Kraus, J. D., *Antennas for all Applications, Third Edition*, R. J. Marhefka (ed), New York: McGraw Hill, 2002.

Appendix 2A

Examples of Industrial Omnidirectional Vertical Antennas

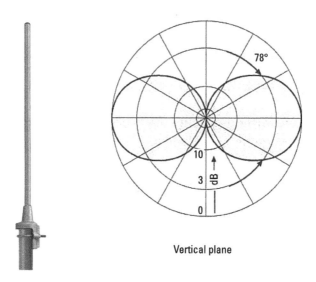

Vertical plane

Figure 2.24   Half-wave antenna, 406–470 MHz.

Technical Characteristics

- Frequency band: 406–470 MHz.
- Vertical polarization.
- Aperture angle: 78°.
- Gain: 2 dBi.
- Impedance: 50 Ω.
- VSWR < 1.5.
- Maximum power: 100 Watts.
- Height: 510 mm.

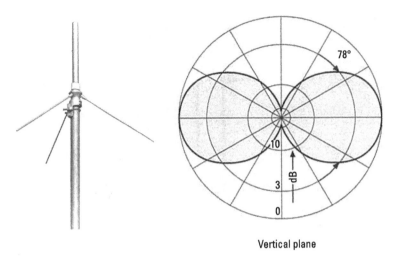

Vertical plane

Figure 2.25 Ground plane dipole antenna, 146–174 MHz.

Technical Characteristics

- Frequency band: 146–174 MHz.
- Vertical polarization.
- Aperture angle: 78°.
- Gain: 2 dBi.
- Impedance: 50 Ω.
- VSWR < 1.5.
- Maximum power: 170 Watts.
- Length:
  - Radiating strand: 422 mm.
  - Radian: 617 mm.

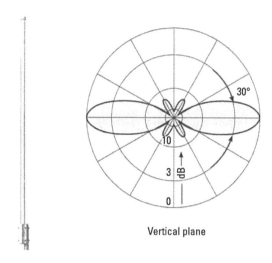

Vertical plane

Figure 2.26   Two-dipole omnidirectional antenna 380–400 MHz.

Technical Characteristics

- Frequency band: 380–400 MHz.
- Vertical polarization.
- Aperture angle: 30°.
- Gain: 5 dBi.
- Impedance: 50 Ω.
- VSWR < 1.5.
- Maximum power: 500 Watts.
- Height: 1612 mm.

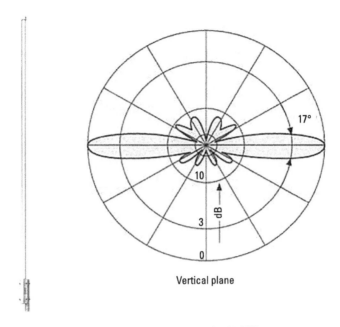

Figure 2.27 Four-dipole omnidirectional antenna, 380–400 MHz.

Technical Characteristics

- Frequency band: 380–400 MHz.
- Vertical polarization.
- Aperture angle: 17°.
- Gain: 7.5 dBi.
- Impedance: 50 Ω.
- VSWR < 1.5.
- Maximum power: 500 Watts.
- Height: 2840 mm.

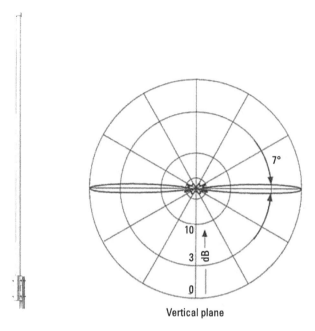

Vertical plane

Figure 2.28   Six-dipole omnidirectional antenna, 870–960 MHz.

Technical Characteristics

- Frequency band: 870–960 MHz.
- Vertical polarization.
- Aperture angle: 11°.
- Gain: 9 dBi.
- Impedance: 50 Ω.
- VSWR < 1.5.
- Maximum power: 150 Watts.
- Height: 3033 mm.

3

Panel and Yagi Antennas

1.1 Panel Antennas

The panel antennas consist of dipoles aligned vertically according to the same principle as the omnidirectional vertical wire antennas (first type), as seen in the previous chapter. However, the dipoles are placed in front of a conductive plane serving as a screen, so as to produce a directive antenna.

These panels present themselves in the form of fiberglass rectangular boxes of low thickness called *radome*. These panels contain the conductive plane and the dipoles. The panels have known an important development in mobile phone networks since the arrival of cellular networks.

They have become indispensable in all operated cellular networks, such as 2G, 3G, 4G, and private mobile networks (PMR) like TETRA, TETRAPOL, and APCO 25. This is because they allow the user to achieve sectored cells with three cells per site (with three panels at 120°) in areas with high traffic.

This new use of the panels has led to many options:

- Panel with n dipoles, vertically polarized, forming a single antenna.
- A vertical polarization panel with a fixed or adjustable electric tilt that constitutes a single antenna.
- A cross-polarization panel (+45°, −45°), forming two antennas with different polarizations, used to make polarization diversity in reception.
- A cross-polarization panel (+45°, −45°) with fixed or adjustable electric tilt that constitutes two antennas with different polarizations. This is used to make polarization diversity in reception.

- A multiband cross-polarization panel (+45°, −45°) with two, three, or four antennas.

- A multiband cross-polarization panel (+45°, −45°) with two, three, or four antennas with fixed or adjustable electric tilt.

In this chapter, we will see only the panel with n dipoles, vertically polarized, constituting of a single antenna. In the second part of this book (*Engineering of Antenna Sites*), we will see all the other options.

1.1.1 Basic Study—Radiation of a Dipole Placed at a Distance d from a Metal Screen

Figure 3.1 shows a dipole A_1 placed at a distance d of a metal screen.

The screen will turn this antenna into a directive antenna; that is, it will have a maximum radiation intensity in a direction, which as we shall see, is perpendicular to the screen. Let axis oz be this maximum propagation direction and let axes ox and oy be the other two axes allowing to form a *direct orthogonal axis system* (see Appendix B). According to the image theory, it is as if the metal screen did not exist and there were two dipoles: A_1 and A_2, the latter being symmetrical to A_1 relative to the screen, and carrying a current in

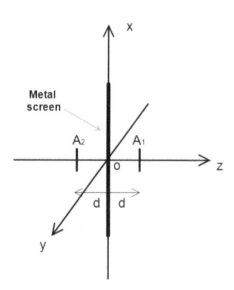

Figure 3.1   A single dipole A_1.

phase opposition. We will calculate successively the radiation pattern of this antenna in the E-plane and in the H-plane.

1.1.1.1 Radiation Pattern in the E-Plane

The E-plane is the plane containing the direction of maximum radiation and the electric fields in that direction (see Chapter 1). It is, therefore, the *xoz* plane, which we have represented separately in Figure 3.2.

As we have explained in Chapter 1, the dipole and its image are radiating from the center. The electric field provided by the dipole A_1 at a point M faraway at the distance r in the direction θ is [see (2.4) in Chapter 2].

$$E_1 = 60 I_M \cos(\pi/2 \, \sin\theta)/r \ \cos\theta e^{j(\omega t - 2\pi r/\lambda)}$$

where θ is, right now, the OM angle with the horizontal axis, and not with the vertical axis as Chapter 2. The electric field provided by the dipole A_2 at the point M is

$$\bar{E}_2 = \bar{E}_1 e^{-j(4\pi d \cos\theta/\lambda + \pi)} e^{j(\omega t - 2\pi r/\lambda)}$$

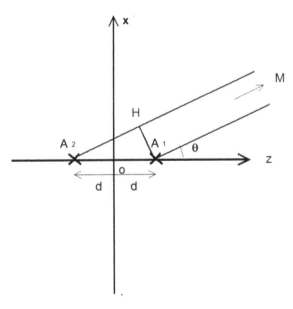

Figure 3.2   E-plane.

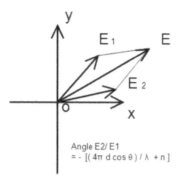

Figure 3.3   Representation in the complex plane.

Because, in addition to the phase shift of π of the current in the antenna image, there is a delay due to the path of the wave emitted by the latter on the distance A2H.

The total field is $\overline{E} = \overline{E}_1 + \overline{E}_2$.

This field can be represented in the complex plane (as seen in Figure 3.3) by the geometrical sum of two vectors $\vec{E}_1$ and $\vec{E}_2$.

The total field is thus in module

$$E = -2E_1 \cos\left[(2\pi d \cos\theta)/\lambda + \pi/2\right]$$

If $d = \lambda/4$, we find

$$E = -2E_1 \cos[\pi/2\ (\cos\theta + 1)] \tag{3.1}$$

Setting $\theta = 0$ in (3.1), we find that the maximum electric field is $E = 2E_1$. Therefore, the direction $\theta = 0$ is the direction of the maximum radiation intensity U_{max}.

So let's calculate now the isotropic gain G_i.

The electric field gain in the direction $\theta = 0$ is equal to 2. The power gain is, therefore, equal to 4 (hence 6 dB), because $W = 1/2\ \varepsilon\ E^2$.

As we lose 3 dB at the antenna input because the input power is split in half on the dipole and on the image (image theory), the antenna gain relative to the dipole is thus 3 dB.

$G_d = 3$ dB therefore, $G_i = 5.16$ dB

Replacing E_1 with its value, we obtain:

$$E = -120 I_M \cos(\pi/2\ \sin\theta)/r \cos\theta\ \cos[\pi/2\ (\cos\theta + 1)]$$

Table 3.1
Characteristic Function

θ	$f(\theta)$
0°	1
15°	0.95
30°	0.8
45°	0.55
60°	0.14
75°	0.08
90°	0

The characteristic function of the field E in the E-plane is thus:

$$F(\theta) = -\left[\cos(\pi/2 \sin\theta)/\cos\theta\right] \cos\left[\pi/2 \left(\cos\theta + 1\right)\right] \qquad (3.2)$$

which allows us to draw the radiation pattern in the E-plane for $d = \lambda/4$ from Table 3.1.

We have varied θ only between 0 and 90°. This is because if we change θ into $-\theta$ in the formula 3, we obtain a symmetric curve relative to the horizontal axis. The radiation pattern in the E-plane (Figure 4.7) is represented in the next paragraph with the pattern in the H-plane.

1.1.1.2 Radiating Pattern in the H-Plane

The H-plane is the *xoz* plane is represented in Figure 3.4.

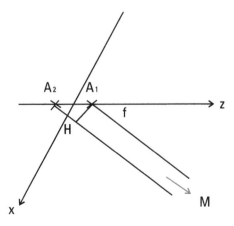

Figure 3.4 H-plane.

In this plane, the electric field emitted by the dipole A_1 at the point M is the same as the field in the E-plane for $\theta = 0$ at the same distance r.

$$\bar{E}_1 = 60I_M/r \; e^{j(\omega t - 2\pi r/\lambda)}$$

The dipole A_2 provides at the point M an electric field E_2 is equal to

$$\bar{E}_2 = \bar{E}_1 e^{-j(4\pi d \cos\varphi/\lambda + \pi)} e^{j(\omega t - 2\pi r/\lambda)}$$

for the same reasons as in the previous case. The phase shift is the same. The resulting field value in module is thus

$$E = -2E_1 \cos[(2\pi d \cos\varphi)/\lambda + \pi/2]$$

For $d = \lambda/4$, we have

$$E = -2E_1 \cos[\pi/2 \, (\cos\varphi + 1)] \qquad (3.3)$$

Replacing E_1 by its value, we obtain

$$E = -120I_M/r \cos[\pi/2 \, (\cos\varphi + 1)]$$

The characteristic function is

$$F(\theta) = \cos[\pi/2 \, (\cos\varphi + 1)]$$

In Table 3.2, we have varied q only from 0° to 90° for the same reasons as previously stated.

We can now plot the two radiation patterns (see Figure 3.5 and 3.6).

Table 3.2
Characteristic Function

ϕ	$f(\phi)$
0°	1
15°	0.99
30°	0.98
45°	0.89
60°	0.7
75°	0.4
90°	0

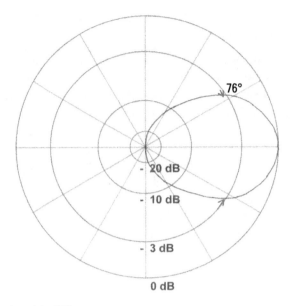

Figure 3.5   E-plane ($d = \lambda/4$).

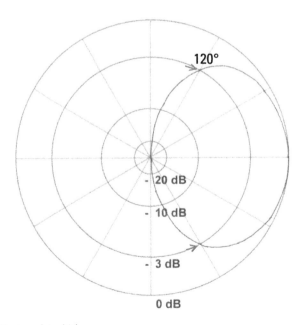

Figure 3.6   H-plane ($d = \lambda/4$).

1.1.2 *N* Dipoles Panel with Vertical Polarization, Constituting a Single Antenna

The dipoles are installed in a vertical linear way inside of a fiberglass flat parallel piped box (commonly known as *radome*), and are at a distance $\lambda/4$ of a metal screen covering one of the inner faces of the box. We will study the four-dipole panel (Figure 3.7), whose principle is the same for a larger number N.

The radome is not represented. The dipoles are fed in the same conditions as the four-dipole omnidirectional antenna (see Figure 2.19 in Chapter 2). The calculation of the total fields provided by the four dipoles at the point M in the E-plane and H-plane would be made the same way as for the four-dipole omnidirectional antenna, examined in Chapter 2. We would find the same array factors, by which the dipole fields at $\lambda/4$ from the panel [see (3.1) and 3.2)] must be multiplie, for the E-plane and the H-plane (see Chapter 2, Section 2.7.4). Therefore, for the E-plane we can write,

$$F = 4\cos(\pi \sin\theta)\cos(2\pi \sin\theta)$$

changing $\cos\theta$ into $\sin\theta$, because θ is the angle of OM with the horizontal axis. For the H-plane, we can write

$$F = 4$$

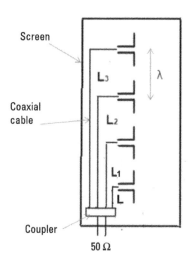

Figure 3.7 Four-dipole panel.

Hence, the values of E_{4d} for the E-plane are

$$E_{4d} = 8E_1 \cos(\pi \sin \theta) \cos(2\pi \sin \theta) \cos[\pi/2 \ (\cos \theta + 1)] \qquad (3.4)$$

where E_1 is the field of a dipole isolated in space in the E-plane. The values of E_{4d} for the H-plane are

$$E_{4d} = 8E_1 \cos[\pi/2 \ (\cos \varphi + 1)] \qquad (3.5)$$

where E_1 is the field modulus of the dipole isolated in space in the H-plane.

1.1.2.1 Gain of the Four-Dipole Panel Antenna

Setting $\theta = 0$ in (3.4), we find that the maximum electric field is $E = 8E_1$. Therefore, the direction $\theta = 0$ is the direction of the maximum radiation intensity U_{max}.

So, let's calculate now the isotropic gain G_i.

The electric field gain in the direction $\theta = 0$ is equal to 8, and therefore has a power gain of 64 (18 dB), because the power is proportional to the square of the electric field ($W = 1/2 \ \varepsilon \ E^2$). As the power input of each dipole is the eighth of the power at the foot of the antenna (-9 dB), because of the 6 dB loss for the distribution of the power input to each dipole and of the 3 dB loss for the image feeding of each dipole, we finally find:

$$G_d = 9 \text{ dB} \qquad G_i = 11.6 \text{ dB}$$

1.1.2.2 Radiation in the E-Plane

The characteristic function in the E-plane is, according to (3.4), replacing E_1 by its value:

$$F(\theta) = [\cos(\pi/2 \sin \theta)/\cos \theta] \cos(\pi \sin \theta) \cos(2\pi \sin \theta) \cos[\pi/2(\cos \theta + 1)]$$

Table 3.3 allows one to draw the radiation pattern in the E-plane.

The radiation pattern in the E-plane is represented in the following paragraph with the pattern in the H-plane.

1.1.2.3 Radiation in the H-Plane

According to (3.5), the radiation pattern in the H-plane is

$$F(\theta) = \cos[\pi/2(\cos \varphi + 1)]$$

Table 3.3
Characteristic Function in the E-Plane of the Four-Dipole Panel

θ	$F(\theta)$	θ	$F(\theta)$
0	1	50	0.03
5	0.8	55	0.13
10	0.38	60	0.22
15	0.04	65	0.17
20	0.7	70	0.12
25	0.4	75	0.07
30	0	80	0.03
35	0.15	85	0.01
40	0.17	90	0
45	0.09		

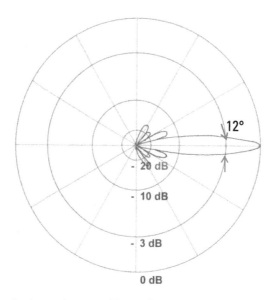

Figure 3.8   Four-dipole panel antenna (E-plane).

The radiation pattern in the H-plane is the same as if there was only a single dipole before the panel. We can thus draw the two patterns E-plane and H-plane of the four-dipole panel antenna (Figures 3.8 and 3.9).

This panel antenna that has an aperture angle equal to 120° in the horizontal plane should be ideal for sectored sites with three sectors of the cellular networks. In fact, for these networks, operators require a coverage of 120° at

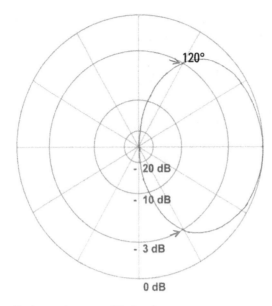

Figure 3.9 Four-dipole panel antenna (H-plane).

−10 dB and not at −3 dB in order to avoid a sectored cell spilling over the adjacent sectored cells. The aperture angle requested by the operators is of the order of 80° and not 120° (see Chapter 6). However, the panel antenna with 120° aperture can be used for other applications.

1.2 Yagi Antennas

The *Yagi antenna* is composed of a dipole, the pilot, a reflector, and as many directors as is necessary to obtain the desired gain (Figure 3.10). All the elements are electrically isolated from their horizontal support.

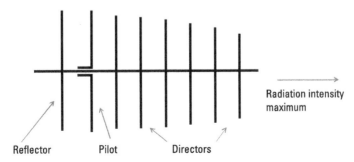

Figure 3.10 Yagi antenna.

It is by the pilot, which is a half-wave dipole, that the antenna is excited. The reflector serves as a metal screen to limit the radiation rearward and favor the radiation forward (maximum radiation intensity). The director elements increase the antenna directivity toward the front.

1.3   Principles of Yagi Antenna

The Yagi antenna is based on the coupling between antennas. This will be discussed in Chapter 4 of this book. The principle relies on two properties.

Figure 3.11 shows a vertical dipole and a conductive wire parallel to the latter, called a *parasite*, at a distance *d*.

1.3.2.1   First Property

The dipole and the parasite are in the same vertical plane and are symmetric to the *oz* axis in that plane. As a result, if we remain in this plane, the vertical components of the $\vec{E}_{inc}$ fields on the parasite—radiated by the dipole at points symmetric to *oz*—provide currents equal in amplitude and in phase opposition in the parasite. Therefore, these currents cancel each other (please see Figure 1.8 in Chapter 1).

The only electric field $\vec{E}_{inc}$ is provided to the parasite by the dipole, which generates current in this parasite is thus the incident field at the center of it.

1.3.2.2   Second Property

It relies on the properties of *good conductors*, especially those of a high-frequency electromagnetic vacuum (see Appendix D). This conductor will therefore not be subjected to any electromagnetic phenomenon, and the tangential electric field to a conductive wire of this quality will thus be zero

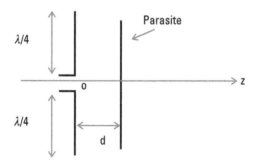

Figure 3.11   Dipole and parasite.

at any point. The good conductor chosen for manufacturing radiating elements is aluminum.

It follows if the parasite is of this quality and if it receives from the dipole a high frequency electric field $\vec{E}_{inc}$ tangential to it in the center (see *first property*), this field then creates a current i, which itself provides a parasite field $\vec{E}_{par}$ that is tangential to it in the center, of the same amplitude as $\vec{E}_{inc}$, but out of phase of π, to cancel it. Using these properties, we will see that if the parasite is spaced at a distance $d = \lambda/4$ away from the pilot, it reflects the electromagnetic wave received from the latter, thus acting as a reflector.

At the contrary, it is possible, with another value of d, to cancel the wave reflected by the parasite and to have only the incident wave along oz. In this case, the parasite acts as a director. Therefore, the director attracts the wave in its direction, hence the name.

1.3.1 Functioning of a Yagi Antenna

1.3.1.1 Reflector

Figure 3.12 shows the pilot and the reflector in a system of axes *oxyz*.

The pilot A_1 is a dipole, (two $\lambda/4$ strands), the reflector has a length of $\lambda/2$ or slightly greater, where λ is the wavelength in the aluminum. The reflector is parallel to the dipole and is spaced a distance $d = \lambda/4$ away from the latter (where λ is the wavelength in the air). As we have explained in Chapter 1, the dipole is radiating from the center.

The reflector is radiating from the center. Indeed, we have seen first propriety (see Section 3.2.1) that the electric field $\vec{E}_{inc}$ provides current in the center of the parasite, which means that the reflector radiates from its center.

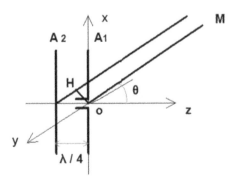

Figure 3.12 Pilot and reflector.

Radiating Pattern in the E-Plane

The electric field provided by the dipole at the point M faraway at the distance r is

$$\bar{E}_1 = E_0 e^{j(\omega t - 2\pi r / \lambda)}$$

with $E_0 = 60I_M \cos(\pi/2 \sin\theta)/r\cos\theta$ [see Chapter 2, (2.3) with θ being angle OM, OZ (Figure 3.12)].

The value of the E_2 field created in M by the reflector A_2 is

$$\bar{E}_2 = \bar{E}_1 e^{j(-\pi/2 - \pi - \pi/2 \cos\theta)}$$

where the delay of $\pi/2$ corresponds to the path of the wave from A_1 to A_2, the delay of π to the second property described above, and the delay of $\pi/2 \cos\theta$ to the path difference between $A_1 M$ and $A_2 M$.

Therefore,

$$E_2 = E_1 e^{j(-3\pi/2 - \pi - \pi/2 \cos\theta))}$$

The total field in M is

$$\bar{E} = \bar{E}_1 + \bar{E}_2 = \bar{E}_1 \left(1 + e^{-j(3\pi/2 + \pi/2 \cos\theta)}\right)$$

This field can be represented in the complex plane, as shown in Figure 3.13, by the geometric sum of two vectors $\vec{E}_1$ and $\vec{E}_2$.

Hence, the modulus of the $\vec{E}$ field

$$E = 2E_1 \cos\pi/4 \, (3 + \cos\theta) \tag{3.6}$$

Setting $\theta = 0$ in (4.5), we find that the maximum electric field is $E = 2E_1$. Therefore, the direction $\theta = 0$ is the direction of the maximum radiation intensity U_{max}.

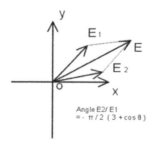

Figure 3.13 Complex plane.

So, let's calculate now the isotropic gain G_i.

The electric field gain in the direction $\theta = 0$ is equal to 2. The power gain is therefore equal to 4 (6 dB), since $W = 1/2\ \varepsilon\ E^2$. As we lose 3 dB at the antenna input because the input power is split in half on the dipole and on the reflector, the antenna gain relative to the dipole is thus 3 dB.

$$G_d = 3 \text{ dB}$$

Therefore,

$$G_i = 5.16 \text{ dB}$$

For $\theta = \pi\ E = 0$, there is no propagation in the direction of the negative z. The reflector acts really as a screen. Replacing E_1 in (3.6) by its value, we obtain:

$$E = 120I_M[\cos(\pi/2\ \sin\theta)\cos\pi/4\ (3+\cos\theta)/r\cos\theta$$

This is the equation of the field provided in M by the association of the dipole and the reflector. The characteristic function in the E-plane is thus

$$F(\theta) = [\cos(\pi/2\sin\theta)\cos\pi/4\ (3+\cos\theta)/\cos\theta$$

Table 3.4 allows one to draw the radiation pattern in the E-plane.

We have varied θ only from 0° to 180°, because the curve is always symmetrical to the axis oz. The radiation pattern in the E-plane will be plotted in the next paragraph with the pattern in the H-plane.

Radiating Pattern in the H-Plane

Figure 3.14 represents a distant point M in the H-plane (xoy-plane).

In this plane, the electric field emitted by the dipole A_1 at the point M is the same as the field in the E-plane for $\theta = 0$ at the same distance r.

$$\bar{E}_1 = 60I_M/re^{j(\omega t-2\pi r/\lambda)}$$

Table 3.4
Calculation Table of $F(\theta)$ (E-Plane)

$\theta°$	$F(\theta)$	$\theta°$	$F(\theta)$
0	1	105	0.11
15	0.95	120	0.38
30	0.80	135	0.14
45	0.60	150	0.09
60	0.38	165	0.03
75	0.18	180	0
90	0		

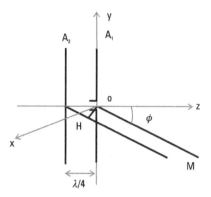

Figure 3.14   H-plane.

As previously shown, the field provided by the parasite (reflector) in M is

$$\bar{E}_2 = \bar{E}_1 e^{-j(3\pi/2+\pi/2 \ \cos\varphi)}$$

and the value of the total field in M is

$$\bar{E}_2 = \bar{E}_1 \left(1 + e^{-j(3\pi/2+\pi/2 \ \cos\varphi)}\right)$$

In plotting the two vectors in the complex plane, we would find as previously, for the E-modulus

$$E = 2E_1 \cos \pi/4 \ (3 + \cos\varphi)$$

The characteristic function is therefore

$$F(\varphi) = \cos \pi/4 \ (3 + \cos\varphi)$$

Table 3.5 allows one to draw the radiation pattern in the H-plane.

Table 3.5
Calculation Table of $F(\varphi)$ (H-Plane)

$\varphi°$	$F(\varphi)$	$\varphi°$	$F(\varphi)$
0	1	105	0.55
15	0.95	120	0.35
30	0.99	135	0.23
45	0.97	150	0.11
60	0.92	165	0.03
75	0.84	180	0
90	0.7		

We can therefore draw Figures 3.15 and 3.16.

In conclusion, we can say that the reflector fulfils its role and gives gain. We can notice that the diagram of the H-plane (Figure 3.16) has the form of a cardioid with an aperture angle of 180°.

The resulting antenna is industrialized for particular applications in which a coverage of 180° is sufficient (some applications in PMR networks).

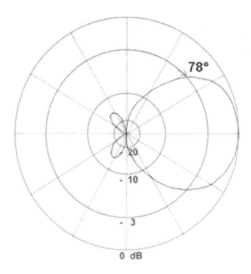

Figure 3.15 E-plane.

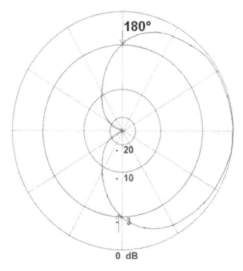

Figure 3.16 H-plane.

It consists of a dipole and an aluminum tube at a distance of $\lambda/4$ from the dipole. This is shown in Figure 3.17.

Technical Specifications

- Frequency band: 68–87.5 MHz.
- Vertical polarization.
- Aperture angle:
  - H Plane: 180°.
  - E Plane: 78°.
- Gain from the dipole: 2 dB.
- Impedance: 50 Ω.
- VSWR < 1.5.
- Maximum power: 230 Watts.

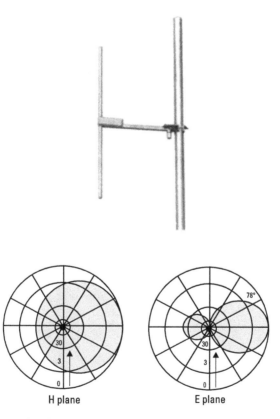

H plane   E plane

Figure 3.17 $\lambda/4$ spacing dipole to tub.

1.3.1.2 Director

Increase of the Antenna Directivity

Let's take the case of the first director adjacent to the pilot. Let's suppose that it is at the distance d of the pilot (two $\lambda/4$ tubes), and let's suppose its length is slightly less than $\lambda/2$, where λ is the wavelength in the pilot and the director (Figure 3.18).

We will see in Section 3.3.2.1 (first property) that the only electric field $\vec{E}_{inc}$ provided to the director by the pilot, which generates current in this parasite is the incident field at the center of the latter.

The result is that the set—pilot and director in the oz direction—is equivalent to an HF line (see Appendix F), merging with oz, if we replace the voltage along the line with the electric field. This two-wire line of length d would be thus powered by a generator (the pilot), and loaded by the input impedance of the director. The director which is isolated in the space is equivalent to an open HF line of length $l/2$. Its impedance is a pure reactance (see Appendix G).

As shown in Figure 3.19, we can plot the representation of the reflection coefficient at the input of this open line on the Smith Chart (see Appendix F). It is represented by the vector $\overrightarrow{\mathbf{OA}}$, whose modulus is 1 ($E_{ref} = E_{inc}$) and which is in the capacitive part of the chart if l is less than $\lambda/2$, because $l/2$ is less than $\lambda/4$ (lower half of the circle $\Gamma = 1$).

The phase shift of $\overline{E}_{ref}$ relative to $\overline{E}_{inc}$ is a phase advance of $2\pi l/\lambda$. The phase difference at a point P, on the left of the pilot, between the field from the pilot (field directed towards the negative z) and the field reflected by the director, is:

$$\Delta\varphi = -4\pi d/\lambda + 2\pi l/\lambda \tag{3.7}$$

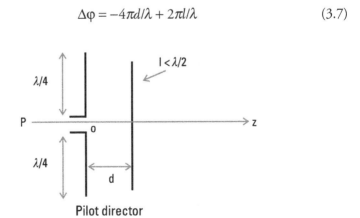

Pilot director

Figure 3.18 Pilot and director.

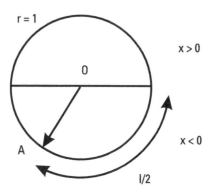

Figure 3.19 Smith chart.

where the distance d is traveled twice: once by the incident wave and once by the wave reflected by the director. If this phase difference is equal to $+/- \pi$, then the electric field at P will be zero because the field from the pilot rearward and the field reflected by the director cancel one another.

On the axis oz there is no field radiated by the pilot in the direction of the negative z, but only in the direction of the director. It can be concluded that, if $\Delta\varphi = +/-\pi$, then the field wave from the pilot on the line is unique, that it is attracted in the direction of the director, and that it carries 75% of the power supplied to the pilot. This is because of the significant coupling between the pilot and the director. The electric field $\vec{E}_{inc}$ of this wave, tangential to the director, will induce in the latter a current i, which in turn, will radiate this parasite in all directions except in the direction of the negative z.

This current will particularly create an electric field wave in the direction oz, which will be attracted by the second director, if we make the same calculation (analogy with a HF line and application of the (3.6) of the phase difference which is linked to l). The field $\vec{E}_{inc}$ will create in the second director current i, which will radiate the latter in all directions, except in that of the negative z, and so on for the other directors.

Reduction of the Input Impedance of the Pilot

The negative effect of the directors is that they reduce the input impedance of the pilot. This is easily demonstrated from the coupling equations between two antennas that we shall see in Chapter 4 of this book.

$$\overline{v}_1 = \overline{z}_{11}\dot{i}_1 + z_{12}\overline{i}_2 \qquad (3.8)$$

$$\overline{v}_2 = \overline{z}_{21}i_1 + z_{22}\overline{i}_2 \tag{3.9}$$

where z_{11} and z_{22} are the proper impedances of the antennas (1) and (2), they correspond to the input impedance z_e of an isolated antenna.

> z_{12} and z_{21} are the mutual impedances between the two antennas. They are no longer dependent on the currents and represent a purely geometrical parameter. By reciprocity, we can write $Z_{12} = Z_{21}$ (see Chapter 4).
>
> v_1 and i_1 represent the voltage and current for the antenna input (antenna 1).
>
> v_2 and i_2 represent the voltage and current for the antenna input (antenna 2).

In the case of the director $v_2 = 0$ and we can replace z_{21} with z_{12}. Therefore, we can write

$$\overline{v}_1 = \overline{z}_{11}i_1 + z_{12}\overline{i}_2$$

$$0 = \overline{z}_{12}i_1 + z_{22}\overline{i}_2$$

The current induced in the director is:

$$\overline{i}_2 = -\overline{z}_{12}/\overline{z}_{22}\ i_2$$

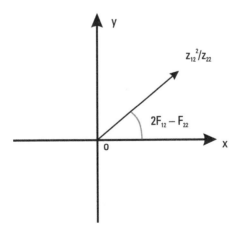

Figure 3.20 Complex plane.

Let's introduce the modulus and the phase of each impedance (see Appendix C).

$$\bar{i}_2 = -(z_{12}e^{j\Phi 12}/z_{22}e^{j\Phi 22})\, i_1 = -i_1 z_{21}/z_{22}e^{j(\Phi 12-\Phi 22)}$$

Therefore, the input impedance of the pilot becomes

$$\bar{Z}_e = v_1/i_1 = z_{11} + z_{12}^2/z_{22}e^{j(\Phi 12-\Phi 22)}$$

Let's draw the vector: $z_{12}^2/z_{22}e^{j(\Phi 12-\Phi 22)}$ in the complex plane (as shown in Figure 3.20).

Let's assume that $\bar{z}_{11} = R_{11} + jX_{11}$ and $\bar{z}_e = R_e + jX_e$.

By equating the real and imaginary parts, we find

$$R_e = R_{11} - z_{12}^2/z_{22}\cos(2\Phi_{12} - \Phi_{22})$$

$$X_e = X_{11} - z_{12}^2/z_{22}\sin(2\Phi_{12} - \Phi_{22})$$

We see that the coupling of the pilot to the first director decreases its input resistance and modifies the tuning of the pilot, which consists of a dipole (the reactive part is modified). The increase in the number of directors will only accentuate this phenomenon.

This is why on a Yagi with high gain, we replace the dipole with a *folded dipole*. Indeed it presents, as we will later see, an input impedance that is four times greater than the dipole. The modification of the tuning is corrected by adjusting the dipole length.

1.3.1.3 Yagi Antenna Design

We have seen the respective functioning of the reflector and directors. The directors radiate in all directions—except in the negative z direction—waves out of phase with each other. This is because of the distance between each director and the length of directors. It will be necessary to add to these radiations those of the pilot-reflector set.

The calculation of the radiation patterns and gain in the *oz* direction is therefore extremely complex and requires the use of a software program.

This software will proceed by successive length adjustments (directors and spaces between them) until it reaches the required characteristics. The more the number of directors is high, the more the gain and directivity will be high.

An example of radiation patterns of a Yagi with ten directors

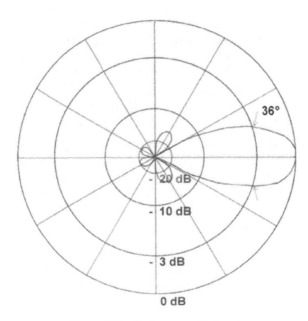

Figure 3.21   Yagi antenna (E-plane).

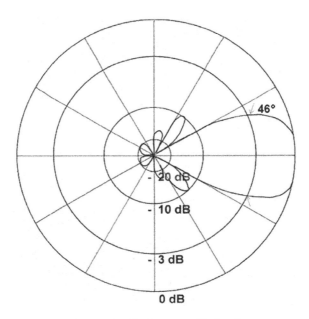

Figure 3.22   Yagi antenna (H-plane).

1.4 Folded Dipole Antenna

A folded dipole antenna is composed of a vertical dipole and a parallel con-ductive wire that are the same length (called parasite), joined by their ends (Figure 3.23).

The distance between the dipole and the parasite is 0.05 λ. Let's apply a voltage V at the dipole input. Given the proximity of the parasite and the dipole, it is as if we had two strongly coupled dipoles, each powered by a volt-age $V/2$ (Figure 3.24).

As there is a current node at the two ends of each dipole, we can connect these ends without changing the characteristics of the two dipoles (Figure 3.25).

This reproduces the schema of the folded dipole (Figure 3.20) with a voltage V distributed as follows:

- $V/2$ for the dipole.
- $V/2$ for the parasite.

The PITOLSKORS Table shown in Chapter 4 gives us $2d = 0.05\lambda$ and $h = 0$ as the coupling resistance and reactance between these two antennas.

$$R_{12} = R_{21} = 73 \ \Omega$$

$$X_{12} = X_{21} = 42 \ \Omega$$

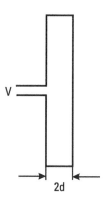

Figure 3.23 Folded dipole.

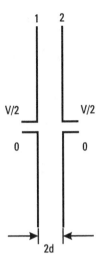

Figure 3.24 Two dipoles.

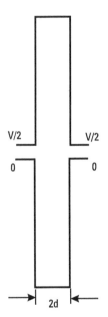

Figure 3.25 Short-circuited ends.

Let's remember the input impedance of the isolated dipole (see Chapter 2).

$$\overline{Z}_e = 73.2\Omega + j42.5\Omega$$

Therefore,

$$\overline{Z}_{12} = \overline{Z}_{21} = \overline{Z}_e$$

Equation (3.8) of the two coupled antennas seen in Section 3.2.2.2 becomes

$$\overline{V}/2 = \overline{Z}_e i_1 + \overline{Z}_e i_2$$

The schema shown in Figure 3.25 is perfectly symmetrical. We can thus write: $i_1 = i_2$.

Hence,

$$\overline{V} = 4\overline{Z}_e i_1$$

The input impedance of the folded dipole is then multiplied by four.

References

[1]   Badoual, R., *Microwaves, Volume 2*, S. Jacquet (ed), France: Masson, 1995.

[2]   Stutzman, W. L., *Antenna Theory and Design, Second Edition*, G. A. Thiele (ed), Hoboken, NJ: Wiley, 1998.

Appendix 3A
Examples of Industrial Directional Antennas

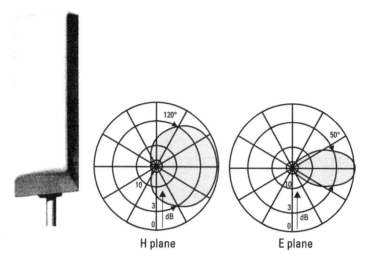

Figure 3.26 Two-dipole antenna, 400–470 MHz.

Technical Specifications

- Frequency band: 400–470 MHz.
- Vertical polarization.
- Aperture angle:
  - H-Plane: 120°.
  - E-Plane: 50°.
- Gain: 5 dBi.
- Impedance: 50 Ω.
- VSWR < 1.5.
- Maximum power: 500 Watts.
- Height: 992 mm.
- Width: 492 mm.
- Depth: 190 mm.

Figure 3.27 Eight-dipole Antenna, 860–960 MHz.

Technical Specifications
- Frequency band: 870–960 MHz.
- Vertical polarization.
- Aperture angle:
  - H-plane: 65°.
  - E-Plane: 13°.
- Gain: 15 dBi.
- Front/Back ratio > 25 dB.
- Impedance: 50 Ω.
- VSWR < 1.3.
- Maximum power: 500 Watts.
- Height: 1294 mm.
- Width: 258 mm.
- Depth: 103 mm.

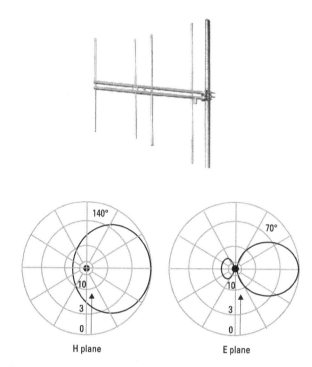

Figure 3.28   Yagi antenna, 68–87.5 MHz.

Technical Specifications

- Frequency band: 68–87.5 MHz.
- Vertical polarization.
- Aperture angle:
  - H-plane: 140°.
  - E-plane: 70°.
- Gain relative to the dipole: 6 dB.
- Impedance: 50 Ω.
- VSWR < 1.5.
- Maximum power: 100 Watts.
- Height: 2380 mm.
- Length: 2030 mm.

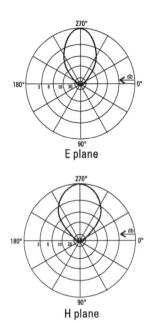

E plane

H plane

Yagi with folded dipole

Figure 3.29   Yagi antenna, 146–174 MHz.

Technical Specifications

- Frequency band: 146–174 MHz.
- Vertical polarization.
- Front/back ratio: 20 dB.
- Aperture angle:
  ○ H-plane: 64°.
  ○ E-plane: 48°.
- Isotropic gain: 12 dB.
- Impedance: 50 Ω.
- VSWR < 1.5.
- Maximum power: 250 Watts.
- Height: 364 mm.
- Length: 800 mm.

Part II:
Antennas Site Engineering

4

Antenna Coupling

4.1 Effect of Coupling between Antennas

The coupling of antennas has two effects. One is harmful, and the other is useful.

4.1.1 The Harmful Effect

With the development of radio communications, the sites are increasingly loaded with antennas. The coupling of antennas results in

- The mismatch of transmitters-receivers due to the lowering of input impedances (see Chapter 3, Section 3.3.1.2).
- Intermodulation products (see Chapter 6).
- Changes in the radiation patterns.

4.1.2 The Useful Effect

The Yagi antennas and the folded dipole antennas are based on the use of coupling between antennas.

4.2 Reciprocity Theorem of Carson

Let's take two dipoles (1 and 2) near one from the other (Figure 4.1) for two examples of the coupling of antennas.

The first experience (on the left) demonstrates the following:

- If v_1 is supplied to the antenna 1 input, the millampere-meter in the antenna 2 input indicates i_2.

The second experience (on the right) demonstrates the following:

- If v_2 is supplied to the antenna 2 input, the millampere-meter in the antenna 1 input indicates i_1.

The Carson theorem states that if $v_2 = v_1 \dots i_2$ is equal to i_1 in amplitude and phase. Therefore, if we call the following ratios *mutual impedance*,

$$Z_{12} = v_1/i_2 \text{ and } Z_{21} = v_2/i_1$$

we can write $Z_{12} = Z_{21}$

4.3 Mutual Impedances

When an omnidirectional vertical antenna A_1, powered, is placed in the vicinity of an omnidirectional vertical antenna A_2, also powered ($2d = \lambda$ for example), then the characteristics of the antennas are modified (Figure 4.2).

This is what is called interaction between two antennas. By analogy with the circuits magnetically coupled, we can write

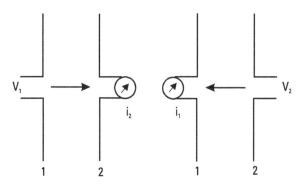

Figure 4.1 Carson theorem.

$$\bar{v}_1 = \bar{z}_{11} i_1 + \bar{z}_{12} i_2$$

$$\bar{v}_2 = \bar{z}_{21} i_1 + \bar{z}_{22} i_2$$

where Z_{11} and Z_{22} are the input impedances of the antennas (1) and (2) when they are isolated. Z_{12} and Z_{21} are the mutual impedances between the two antennas seen above. They no longer depend on the currents and represent a purely geometrical parameter. The calculations of $Z_{12} = R_{12} + jZ_{12}$ have been made by many authors.

In Figure 4.3 and Table 4.1, we give the results obtained by Pistolkors for two $\lambda/2$ dipoles, depending on their relative distance in the horizontal plane and in the vertical plane.

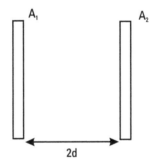

Figure 4.2   Interaction between two antennas.

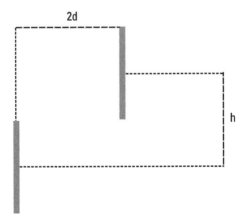

Figure 4.3   Relative position of two dipoles.

Table 4.1

PISTOLKORS Table

$\dfrac{2d}{\lambda}$	$h = 0$		$h = \dfrac{\lambda}{2}$		$h = \lambda$	
	$R_{12}(\Omega)$	$X_{12}(\Omega)$	$R_{12}(\Omega)$	$X_{12}(\Omega)$	$R_{12}(\Omega)$	$X_{12}(\Omega)$
0.00	73.3	42.2	25.1	19.8	−4.1	0.7
0.05	71.7	24.5	19.7	10.8	−4.1	−0.6
0.10	67.3	7.6	17.1	3.1	−4.1	−0.4
0.15	60.4	−7.1	13.7	−3.6	−4.1	0.0
0.20	51.4	−19.1	9.7	−8.6	−4.0	0.5
0.25	40.8	−28.3	4.8	−12.0	−3.9	1.1
0.30	29.2	−34.6	−0.7	−13.1	−3.5	1.7
0.35	17.5	−37.4	−5.7	−13.4	−3.1	2.4
0.40	6.0	−37.6	−10.5	−11.7	−2.4	3.1
0.45	−3.4	−34.8	−13.9	−8.7	−1.6	3.7
0.50	−12.5	−29.9	−16.8	−4.8	−0.7	4.1
0.55	−19.0	−23.4	−18.1	−0.2	0.4	4.2
0.60	−23.2	−16.5	−17.9	4.2	1.5	4.2
0.65	−25.2	−8.0	−16.5	9.1	2.6	3.8
0.70	−24.6	−0.1	−13.7	13.1	3.6	3.1
0.75	−22.5	6.6	−10.0	16.1	4.4	2.3
.080	−18.5	12.2	−5.6	18.0	5.1	1.1
0.85	−13.0	16.3	−0.8	18.7	5.3	−0.2
0.90	−7.5	18.5	4.0	18.0	5.3	−1.6
0.95	−1.5	19.0	8.5	16.1	4.7	−3.0
1.00	4.0	17.3	12.0	13.1	4.0	−4.2
1.10	12.3	11.2	16.2	5.0	1.5	−5.9
1.20	15.2	2.0	15.3	−3.9	−1.3	−6.0
1.30	12.6	−6.7	10.7	−10.8	−4.5	−4.5
1.40	6.0	−11.8	3.2	−14.0	−6.5	−1.6
1.50	−1.9	−12.6	−4.7	−12.8	−6.0	1.9
1.60	−8.1	−8.4	−10.0	−7.8	−4.3	4.8
1.70	−10.9	−2.0	−11.9	−1.1	−1.2	6.3
1.80	−9.1	4.5	−9.6	5.3	2.5	6.0
1.90	−5.4	8.6	−5.3	9.3	5.1	5.8
2.00	1.1	9.2	0.9	9.7	6.2	0.4
2.10	6.0	6.7	5.9	7.3	5.4	−3.0
2.20	8.2	1.8	8.5	2.6	2.8	−5.3
2.30	7.5	−3.3	7.9	−2.3	−0.5	−5.9
2.40	4.0	−6.8	4.9	−6.2	−3.6	−4.6
2.50	0.9	−7.3	0.4	−6.7	−5.4	−1.8
2.60	−4.8	−5.2	−3.6	−5.7	−5.4	1.6
2.70	−6.7	−1.7	−5.8	−2.8	−3.6	4.1
2.80	−6.3	2.6	−5.5	1.2	−0.7	5.4
2.90	−3.4	5.6	−4.2	4.3	2.4	4.7
3.00	−0.3	5.8	−1.6	5.2	4.2	2.3

4.4 Recommended Distance between Two Antennas on a Pylon and on a Terrace

Table 4.1 demonstrates that when the two antennas are aligned vertically on a pylon ($2d = 0$) and that the distance h is equal to λ, the mutual impedance is negligible.

$$Z_{12} = -4.1\ \Omega + j\ 0.7\ \Omega$$

When the two antennas are in the same horizontal plane ($h = 0$), on a terrace of a building, for example, and that $2d$ is equal to $3\ \lambda$, the mutual impedance is negligible.

$$Z_{12} = -0.3\ \Omega + j\ 5.8\ \Omega$$

Table 4.1 shows that the coupling between the dipoles is negligible:

- On a pylon, on condition that they are aligned vertically with a minimum distance of λ between antennas.
- On a building terrace, provided that they are installed at a minimum distance of 3λ.

It may be noted that the first rule applies in the construction of all omnidirectional vertical antennas with gain (candle antennas are discussed in Chapter 2) and panel-type directional antennas (discussed in Chapter 3). The dipoles that compose them are separated by λ. Generally, the two rules must apply to the omnidirectional vertical antennas themselves.

For the panel antennas, the problem does not arise. This is because in the cellular network (main application), panel antennas are always used in groups of three in the same horizontal plane and distant in azimuth of 120°. They are therefore perfectly isolated.

For the parabolic antennas (see Chapter 10), their high directivity makes their isolation very easy. Precautions are to be taken as to the Yagi antennas:

- On a pylon, the rule of minimum spacing of λ with the other antennas is to be applied.
- On a terrace, the rule of 3λ must be applied, but we must also ensure evidently that the main lobe of the Yagi antennas is not pointed on a nearby antenna.

Reference

[1]   Eyraud, L., G. Grange, and H. Ohanessian, *Theory and Techniques of Antennas*, Paris, France: Vuibert, 1973.

5

Antenna Coupling with Mast and Pylon

5.1 Influence of the Pylon or Mast on the Radiation Pattern of an Antenna in the Horizontal Plane

The influence of the pylon or mast on the radiation pattern of an antenna in the horizontal plane depends on:

- The kind of support (pylon or tubular mast);
- The antenna distance;
- The section of the pylon or mast.

5.1.1 Tubular Mast

The tubular mast is used to install antennas on a terrace. We will take four examples (Figures 5.1, 5.2, 5.3, and 5.4) with masts of different diameters (60, 160, 250, and 600 mm) and with different distances d between mast and antenna.

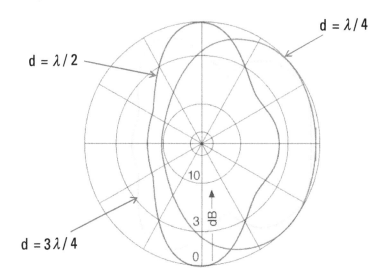

Figure 5.1 H-plane, mast diameter 60 mm.

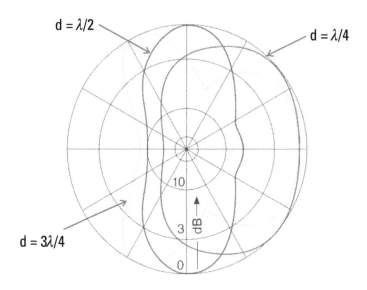

Figure 5.2 H-plane, mast diameter 160 mm.

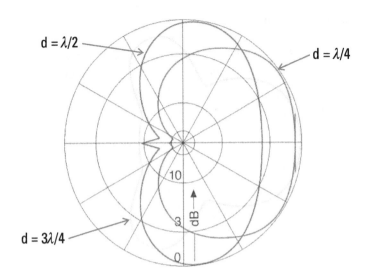

Figure 5.3 H-plane, mast diameter 250 mm.

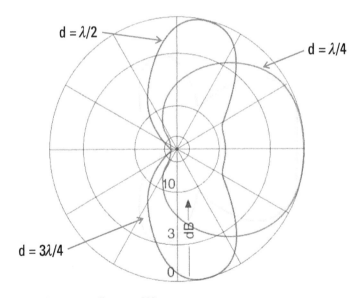

Figure 5.4 H-plane, mast diameter 600 mm.

The analysis of these diagrams made experimentally shows that

- For $d = \lambda/4$, we find a cardioid, which is normal because we are in the same situation as the pilot and reflector of the Yagi (see Chapter 3).
- In this case the mast provides a high directivity to the antenna and a 3-dB gain relative to the dipole.
- This directivity is less sensitive when the diameter of the tubular mast decreases.
- The omnidirectional antenna is obtained for a distance d from the antenna to the mast of 3/4 λ or more and for a mast diameter ranging from 60 mm to a maximum of 100 mm. In this case the maximum variation of the field is 6 dB (by extrapolating for 100 mm), which can be compensated by increasing the gain of the antenna, if necessary.

For reasons of mechanical strength, as the optimum diameter is small, we must not mount more than two or three omnidirectional antennas on a tubular mast installed on a terrace. For the directional antennas such as the three cellular panels, there is no limit to the diameter due to the directivity of the antennas.

5.1.2 Pylon

The pylons make it possible to install a large number of antennas on sites, as we can see as when we leave the cities and cross the rural areas. We can also see that, on these pylons, the omnidirectional antennas are always installed on the smallest section of the pylon. This observation is not without reason.

Here is a common example, with the parameters a and d shown in Figure 5.5, for the pylons of equilateral triangular section, which are the most widespread:

Example : $a = 500$ mm (Figure 5.6).

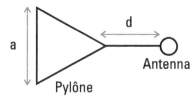

Figure 5.5 Pylon with equilateral triangular section.

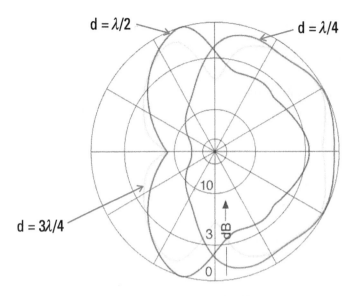

Figure 5.6 H-plane section of the triangular pylon, side 500 mm.

This radiation pattern shows that we find anew the cardioid for I = $\lambda/4$. The pattern is the most omnidirectional for $d = 3/4\ \lambda$ or more.

The maximum variation of the field relative to the field of the preferred direction 0° is only 8 dB (which is the acceptable limit) because we can reduce this gap by increasing the gain of the antenna if necessary.

By analogy with the tubular masts, we can understand that if the side is larger than 50 cm the field variation with the azimuth will be more important.

5.2 Conclusion on the Pylons

This pattern shows that if we want to install an omnidirectional antenna on a pylon, of triangular section, the maximum length of the side can only be 50 cm, and providing we place the antenna at a minimum distance $d = 3\ \lambda/4$.

This short length does not prevent us from the possibility of having great pylons. Indeed, we just have to choose freestanding elements, of triangular section, broken down into elements of decreasing section from the bottom, to have a good stability of the pylon. The top elements can, under these conditions, have a side length of 50 cm, allowing the installation of omnidirectional antennas.

The parabolic, panel and Yagi antennas can be installed lower on the elements of the larger section, since their directivity makes them insensitive to the width of the pylon.

Example: 18m Pylon Leclerc Antennas (Figure 5.7).
It is broken down into six elements of 3m long, of different triangular sections:

- 450 mm side for the two top elements;
- 750 mm side for the three lower elements;
- A reducing central element of 750 to 450 mm side.

The omnidirectional antennas will be installed on the top two elements; that is to say, over a length of 6m, and the directional antennas below.

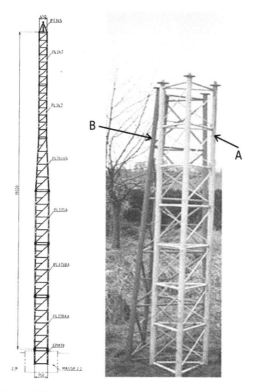

Figure 5.7   18m Pylon. (a) Element of 3m–450 mm side, upper part. (b) Reducing element of 3m–750 mm/450 mm.

6

Cellular Networks: Antenna Tilt, Polarization Diversity, and Multiband on Panels

6.1 Panel Antenna with Vertical Polarization, *n* Dipoles, and Mechanic Fixed Tilt, Electric Fixed Tilt, or Electric Adjustable Tilt

In cellular networks with heavy traffic (generally in urban areas) the cells are small and sectored, with three panels at 120° from each other. It is important to focus the energy transmitted by the transmitters of each sectored cell for two reasons:

- To reduce the transmitter power to cover only the sectored cell;
- To decrease the risks of interference with the sectored cells using the same frequencies (for specialists, to meet the C/I ratio of the network: C/I = ratio carrier over interference).

However, the pattern in the E-plane in Figure 6.1 shows that the radiation is directed toward the horizon. Instead of remaining horizontal, if we want to concentrate the energy in the sectored cell only, the beam must be inclined downwards. This is what is called the *tilt*. Figures 6.2 and 6.3 give an example of two tilts.

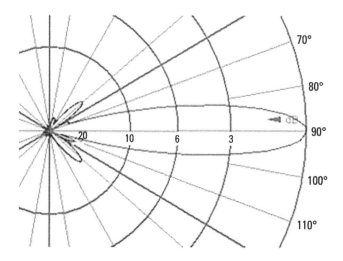

Figure 6.1 E-plane without tilt.

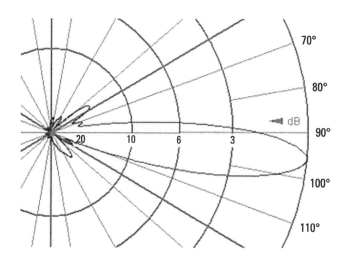

Figure 6.2 E-plane, tilt 5°.

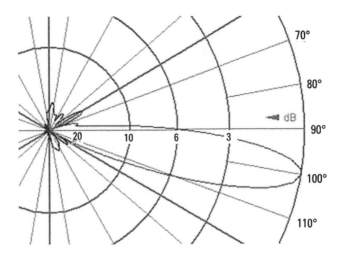

Figure 6.3   E-plane, tilt 10°.

There are two kinds of tilt:

- Mechanical tilt;
- Electric tilt.

6.1.1 Mechanical Tilt

A panel with a mechanical tilt is represented in Figure 6.4.

A mechanical kit makes it possible to incline the panel of the desired number of degrees. The characteristics of a mechanical tilt are

- The tilt is actually realized only in the direction of maximum gain.
- At +/− 90° in azimuth, the tilt is zero.
- For the other azimuths the tilt varies between the chosen tilt and 0, which distorts the radiation pattern (see the example in Figure 6.5 for three different tilts). These are disadvantages that the electric tilt does not have.

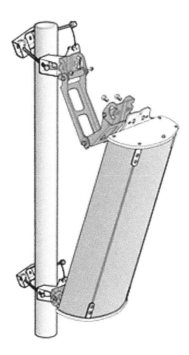

Figure 6.4 Mechanical tilt.

6.1.2 Electric Tilt

6.1.2.1 Electric Tilt Principle

We will start by explaining the principle on two dipoles (Figure 6.6). We will demonstrate that it is possible to perform a tilt of θ° on the radiation pattern of these two dipoles in the E-plane, by phase-shifting of Φ the excitation of a dipole with respect to the other.

Let us suppose the two dipoles are excited by two phase-shifted currents of Φ angle.

The field E_1 created by the dipole A_1 in a remote point M in the ground direction has the value :

$$\overline{E}_1 = E_0 e^{j(\omega t - 2\pi r/\lambda)}$$

with:

$$E_0 = 60 I_M \cos(\pi/2 \sin\theta)/r \cos\theta$$

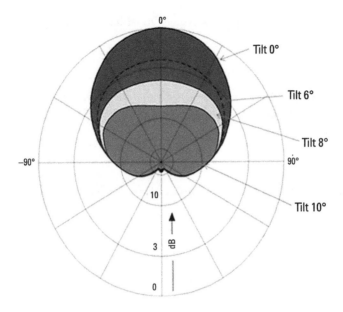

Figure 6.5 H-plane; examples of mechanical tilts.

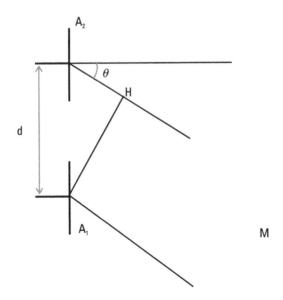

Figure 6.6 Two dipoles.

The field E_2 created by the dipole A_2 in M has the value:

$$\overline{E}_2 = E_0 e^{j(\omega t - 2\pi r/\lambda - 2\pi d \cos\theta/\lambda + \Phi)}$$

$2\pi d \cos\theta/\lambda$ being a delay of phase and Φ the out of phase of the excitation currents. Therefore, the total field in M is:

$$\overline{E} = E_0 e^{j(\omega t - 2\pi r/\lambda)}\left[1 + e^{j(-2\pi d \cos\theta/\lambda + \Phi)}\right]$$

which can be represented in the complex plane as seen in Figure 6.7.

The total field can thus be written:

$$E = 2E_0 \cos(\pi d \sin\theta/\lambda - \Phi/2) \qquad (6.1)$$

The direction θ of the tilt will be the direction of maximum propagation in the E-plane, if $E = 2E_0$; that is to say, $\overline{E}_1$ equal to $\overline{E}_2$ in amplitude and in phase. Therefore, $\pi d \sin\theta/\lambda - \Phi/2 = 0$ (see (6.1)).

Let us assume that Φ is positive, which means A_2 is excited by a current in advance of phase with respect to the current of A_1. The tilt of θ will thus be obtained for a phase shift of:

$$\Phi = 2\pi d \sin\theta/\lambda \qquad (6.2)$$

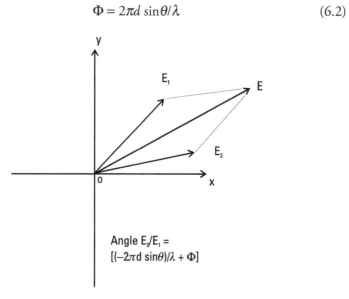

Figure 6.7   Representation in the complex plane.

6.1.2.2 Shift of Phase Realization

The manufacturers propose two systems:

- A shift of phase by coaxial cable length between the source and the dipoles;
- A phase shift circuit between the HF source and the dipoles.

The first system is fixed and cannot be modified easily on the site. The second system is adjustable on the site and can be remote controlled.

6.1.2.3 Shift of Phase by Coaxial Cable Length

Let us make the application with four dipoles. The antenna assembly will be made according to that shown in Figure 6.8.

The lengths of coaxial cables L, L_1, L_2, L_3 (see Chapter 2, Figure 2.19) will have to follow these rules:

L becomes: $L + 3l$

L_1 becomes: $L + \lambda + 2l$

L_2 becomes: $L + 2\lambda + l$

L_3 becomes: $L + 3\lambda$

with: $l = \lambda\ \Phi/2\pi$

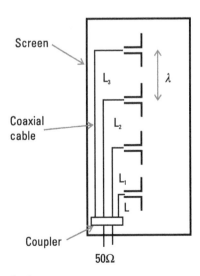

Figure 6.8 θ tilt for four dipoles.

which can be written after replacing Φ by its value (6.2) and d by λ:

$$l = d\,\sin\theta = \lambda\,\sin\theta \qquad (6.3)$$

The principle would be the same for panels with more than four dipoles. Let us show a tilt of 12° for a panel with five dipoles spaced each other of λ (Figure 6.9). We would find $l = 0.208\lambda$, which would correspond to a phase shift between each dipole of $\Phi = 75°$.

6.1.2.4 Phase Shift Circuit

The advantage of a phase shift circuit is that it is possible to adjust the tilt on the antenna site or at distance by a remote control. It is difficult to have information on this system because each manufacturer has taken a patent. However, let us present the principle. At the input of each dipole is installed an adjustable shift phase circuit controlled by a microprocessor. According to Figure 6.9 the phase shift of each dipole would be from the top of the antenna 0, ϕ, 2ϕ, 3ϕ, 4ϕ, and so on.

For that, the microprocessor knowing the tilt θ calculates ϕ by (6.2):

$$\Phi = 2\pi d\,\sin\theta/\lambda = 2\pi\,\sin\theta$$

(because $d = \lambda$), and adjust each phase shift.

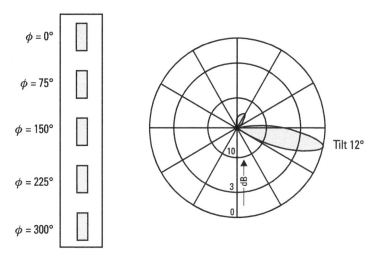

Figure 6.9 Example for a panel with five dipoles.

6.1.2.5 Electric Tilt Benefits

An electric tilt has the following advantages:

- The tilt angle is constant in all directions.
- The shape of the radiation pattern in the horizontal plane (H-plane) remains constant.

It is interesting to compare the radiation pattern of the mechanical tilt with the electric tilt. The benefits of the electric tilt are clearly visible in the radiation pattern in the horizontal plane represented hereafter for three tilts (Figure 6.10). Manufacturers offer both fixed tilts and adjustable tilts.

6.2 Antenna Panel for Polarization Diversity in Reception

6.2.1 Space Diversity and Polarization Diversity

Both space and polarization diversity systems are mainly used in cities, but also in some rural areas, because of the *multipath* waves caused by reflections on buildings (Figure 6.11).

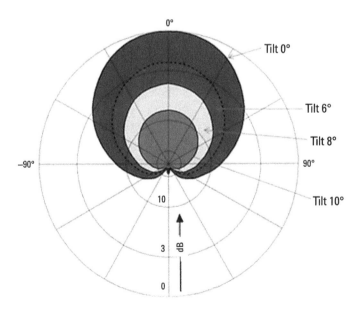

Figure 6.10 H-plane: examples of an electric tilt.

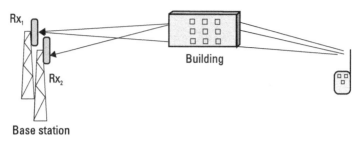

Figure 6.11 Multipaths.

- The signals received by the base station include a direct wave and a certain number of reflected waves according to the density of buildings;
- The reflected waves differ in amplitude, phase and polarization;
- The sum of the signals received by the receiver of the base station varies a lot over short distances.

The first solution to limit these variations is to have two receivers on the base station, with antennas having the same polarization (vertical), separated by 10λ in the horizontal plane, and to choose the signal having the best signal/noise ratio. This is what is called space diversity. The second solution is to have antennas with orthogonal polarizations for the two receivers. This is what is called polarization diversity.

For the first type of diversity, antenna panels with vertical polarization (as seen in Chapter 3) will be used. In the second case, only one panel with two antennas will be used with the dipoles being perpendicular to each other.

Space diversity is more difficult to implement than polarization diversity on a pylon because it would be very difficult to put two antennas spaced 10λ in the horizontal plane. This is only possible on a terrace. This is why polarization diversity is the most used.

6.2.2 Principle of the Panel Comprising Two Antennas with Orthogonal Polarization

Two types of orthogonal polarizations exist:

1. One antenna with vertical polarization and one with horizontal polarization;
2. One antenna with polarization at $+45°$ relative to the vertical and one with a $-45°$ relative to the vertical.

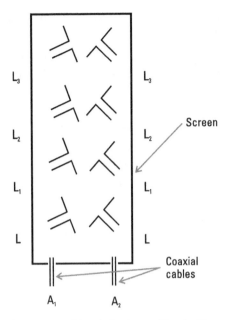

Figure 6.12   Two panel antennas with polarizations +45 and −45.

The second type of orthogonal polarization is used the most because it gives results equivalent to the first type in urban areas, but better results in rural areas. We will therefore only deal with the latter.

We have represented in Figure 6.12 a panel comprising two four-dipole antennas with polarizations +45° and −45° without the fiberglass radome. The dipoles of each antenna are fed by coaxial cables and a coupler, as seen in Figure 6.8. The angle of 90° between two dipoles of the same rank provides a decoupling between the two antennas of at least 30 dB. The diversity of polarization can thus be perfectly achieved. It is also possible to have this type of antenna with an electric tilt. For that L, L_1, L_2, and L_3 must be adjusted on each antenna as in the Section 6.1.2.3.

6.3 Multiband Panel Antennas

A multiband panel antenna is composed of two or three antennas in the same radome so the antenna can transmit and receive two or three bands simultaneously. Antennas of different frequency bands are placed vertically in the radome on top of each other to be properly isolated from each other (see Chapter 4). Examples include:

- The dual band panel 880–960 MHz, 1710–1880 MHz used for the two GSM bands allocated by CEPT (see Appendix A).
- The triple band panel 880–960 MHz, 1710–1880 MHz, 1920–2170 MHz used for the two GSM bands and for the UMTS band (see Appendix A).

These antennas can have the following options:

- Fixed tilt or electric adjustable tilt;
- Orthogonal polarizations (+45°/−45°) to make diversity in reception.

Note that antenna manufacturers also provide panel antennas that can be adjusted according to order, on different bands very close to each other. This type of antenna is completely different than those described above.

Appendix 6A

Examples of Industrial Antennas

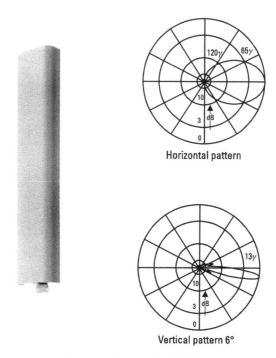

Horizontal pattern

Vertical pattern 6°

Figure 6.13   Panel antenna with eight dipoles 860–960 MHz, a single band, and vertical polarization fixed electrical tilt 6°.

Technical Specifications

- Frequency band: 870–960 MHz.
- Vertical polarization.
- Aperture angle:
  - H-plane, 65°.
  - E-plane, 13°.
- Gain: 15.5 dBi.
- Fixed electrical tilt: 6°.
- Front/back ratio > 25 dB.
- Impedance: 50 Ω.
- VSWR < 1.3.
- Maximum power: 500W.
- Height: 1916 mm.
- Width: 262 mm.
- Depth: 13 mm.

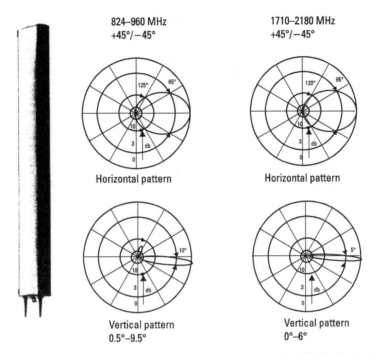

Figure 6.14   Dual band panel antenna 824–960 MHz and 1710–2180 MHz, dual polarization +45°/−45°, adjustable electrical tilt 0.5°–9.5° and 0°–6°.

Table 6A.1
Technical Specifications

Frequency band	824–960 MHz	1710–2180 MHz
Polarization	+45°, −45°	+45°, −45°
Gain	2 × 15 dBi	2 × 18 dBi

Table 6A.2
H Plane Pattern

Aperture angle	65°	65°
Front/back ratio	> 25 dB	> 25 dB

Table 6A.3

E Plane Pattern

Aperture angle	10°	5°
Adjustable electrical tilt	0° to 10°	0° to 6°
Impedance	50 Ω	50 Ω
VSWR < 1.5		
Isolation intersystem	> 30 dB	> 30 dB
Maximum power	250W	250W
Height	1916 mm	1916 mm
Width	262 mm	262 mm
Depth	139 mm	139 mm

7

Filters

7.1 Filter Role

The filter is used in the coupling systems in the duplexers. It is also used to protect against spurious frequencies. The role of the filter is to select a frequency band and reject the other frequencies of the spectrum. Four types of filters are shown in Figure 7.1.

7.1.1 Band-Pass Filter

The ideal band-pass filter passes a frequency band and rejects all other frequencies. The template of the band-pass filter is shown in Figure 7.2.

7.1.2 Band-Stop Filter and Notch Filter

The ideal band-stop filter rejects a frequency band and passes all the other frequencies. The ideal notch filter rejects one frequency and passes all the other frequencies.

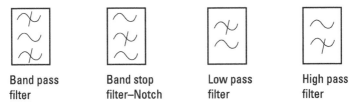

Band pass
filter

Band stop
filter–Notch

Low pass
filter

High pass
filter

Figure 7.1   Representation of four different types of filters.

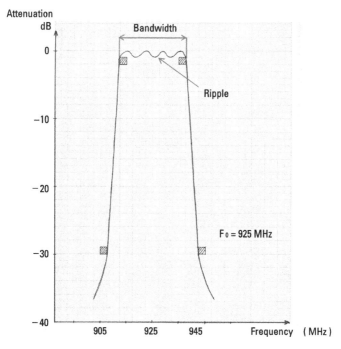

Figure 7.2 The template of a band-pass filter.

7.1.3 Low-Pass Filter

The ideal low-pass filter passes all the frequencies below a frequency called the cutoff frequency and rejects all the frequencies above the latter.

7.1.4 High-Pass Filter

The ideal high-pass filter passes all the frequencies above a frequency called the cutoff frequency and rejects all the frequencies below the latter.

7.2 Filter Characteristics

The characteristics of all filters are:

- The bandwidth.
- The band ripple.
- The insertion loss in the bandwidth.
- The attenuation of out-of-band frequencies.

All these characteristics are represented in a *template* (Figure 7.2 shows an example).

The four specification points define the template of the filter.

7.3 Filter Structures

The filter calculation is performed by a synthesis method, the basic element of which is the low-pass filter. From this, we can also calculate the other types of filters.

The mathematical laws used to calculate these filters are those of Butterworth and Tchebyscheff.

Figures 7.3, 7.4, and 7.5 provide three examples of different filters.

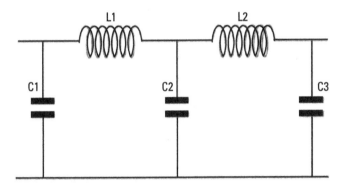

Figure 7.3   Low-pass filter.

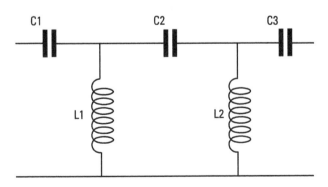

Figure 7.4   High-pass filter.

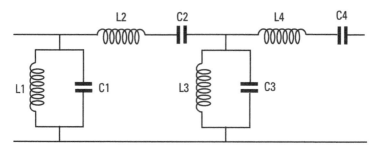

Figure 7.5   Band-pass filter with four resonators.

7.4 Filter Technology

There are three technologies used for the filters according to their frequencies:

- The LC technology (inductance and capacitance) in the lower part of the VHF (bands 34–41 MHz and 68–88 MHz). This is called the *discrete elements* technology.
- The *microstrip lines* technology in the bands VHF (151–162 MHz) and UHF (300–3000 MHz).
- The technology with coaxial cavities in the VHF and UHF bands.

7.4.1 L.C. Technology

The LC technology is based on the use of inductors (L) and capacitor (C) of the trade up to 10 MHz. As mentioned in Appendix G in the section about on microstrip lines over 10 MHz, these components and the conducting wires that connect them together no longer have the same electrical characteristics. This is because of parasitic elements (resistance, inductor, capacity) that are no longer negligible at those frequencies.

However, with a lot of precautions (such as air inductors made with metal wire of significant section, or capacitors with wide and flat metal outputs with very short connections between each component and with a good ground plane) we can use this technology up to frequencies of about 150 MHz.

7.4.2 Microstrip-Line Technology

This technology uses inductors and capacitors made with microstrip lines that have well-defined characteristics for frequency ranges of use in the VHF, UHF bands, and beyond (see Appendix G). By using these components, we have a first microstrip-line technology.

A second technology is that of the coupled lines, which also uses microstrip lines, to make band-pass filters comprising of resonators $\lambda/4$ or $\lambda/2$ (see Appendix G).

7.4.3 Coaxial Cavity Technology

This is the most commonly used technology because it allows for the fabrication of filters with greater selectivity.

7.4.3.1 Principle of the Coaxial-Cavity

The coaxial cavity uses one of the properties of the short-circuited quarter-wave line, the one equivalent to a parallel resonant circuit at resonance (see Appendix G, Section G.5). The principle of this cavity is represented in Figure 7.6.

A coaxial cavity is a cylindrical cavity with an inner metal tube of smaller diameter that is welded to the center of one of the two faces. It is equivalent to a coaxial line, the core being of which is composed of the inner tube. The cavity thus formed is a waveguide (see Appendix H), and is excited by a small antenna or a coupling loop introduced at the input.

An electromagnetic wave propagates in the *fundamental mode* TE_{10} within the cavity in the direction of the cavity axis, and resonates the coaxial line that is terminated by a short circuit (Face AB). The length l of the tube is slightly smaller than $\lambda/4$. This is because the wavelength in a waveguide in TE_{10} mode is greater than λ (speed of the electromagnetic wave greater than the speed in the air; see Appendix H).

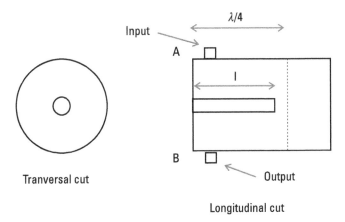

Figure 7.6 Principle of the coaxial cavity.

7.4.3.2 Performing a Coaxial Cavity

Figure 7.7 gives an example of how to fabricate a coaxial cavity with its frequency setting device, consisting of a movable piston, which enables lengthening or shortening the length of the tube.

7.4.4 Comparison of the Coaxial-Cavity and Microstrip-Line Technologies

The cavities, like the microstrip lines, are available in the VHF and UHF bands, but they have the advantage over the latter. This is due to having their quality factor Q well above those of the resonators made with microstrip lines or LC components. The quality factor of a resonant circuit is

$$Q = F_0/\Delta F$$

where ΔF is the bandwidth and F_0 is the resonance frequency. In the VHF and UHF bands, the Q of the cavity is of about 15,000 and even higher. The

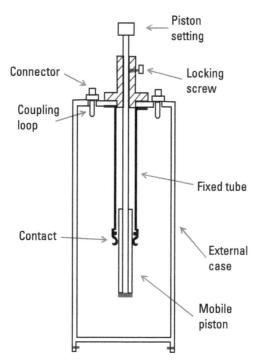

Figure 7.7   Example of coaxial cavity (transversal cut).

choice for the filters between the microstrip lines and the cavities will depend on the desired selectivity.

7.5 Performing a Band-Pass Filter with Coaxial Cavities

The coaxial cavity is equivalent to a parallel resonant circuit. In Figure 7.5, which is a diagram of a band-pass filter, it can therefore replace the parallel resonant circuits. To achieve the series resonant circuits, we use the following property: if we put a parallel resonant circuit between two quarter-wave lines, we will obtain a series resonant circuit. This property comes from the formula giving the input impedance of a quarter-wave line loaded by a pure resistance Z_R (see Appendix G).

$$Z_x = Z_0^2 / Z_r$$

which shows that there is an impedance inversion. We thus obtain Figure 7.8.

The two resonators, L_1C_1 and L_3C_3, are cavities. The two series resonators, L_2C_2 and L_4C_4 (shown in Figure 7.5), are replaced by the two cavities, $L'_2C'_2$, $L'_4C'_4$, and by coaxial links of length $\lambda/4$ on each side of these cavities.

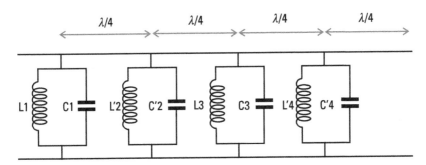

Figure 7.8 Band-pass filter with four resonators.

Reference

[1]   Piette, B., *Multicouplers and Filters VHF/UHF*, France: Lavoisier, 2007.

Appendix 7A

Examples of Industrial Filter

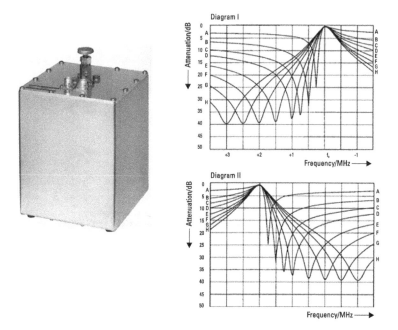

Figure 7.9   Notch filter, 380–470 MHz.

Technology: High Q Coaxial Cavity

Figure 7.9 shows a notch filter.

Technical Specifications

- Frequency band: 380 MHz–470 MHz.
- Impedance: 50 Ω.
- Insertion loss: 0.5 dB
- Power input < 200 Watts.
- VSWR < 1.5.
- Dimensions: 190 mm × 350 mm × 190 mm.

Application

Spurious protection at the receiver input. The cavity is adjustable to attenuate strongly any spurious frequency very close to the operational frequency f_0.

Technology

The band-pass filter shown in Figure 7.10 is an application of Section 7.5.

Technical Specifications

- Frequency band: 1710–1880 MHz.
- Insertion loss < 0.3 dB.
- VSWR < 1.2.
- Impedance: 50 Ω.
- Power input < 500 Watts.
- Dimensions: 210 mm × 253 mm × 110 mm.

Application

Receiver protection against spurious on an antenna site with multiple transceivers.

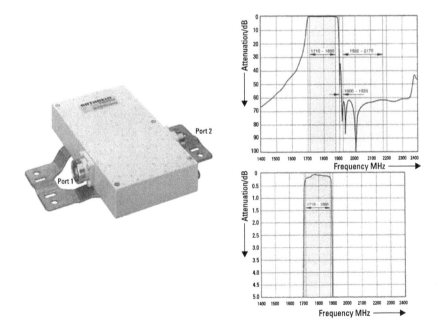

Figure 7.10   Band-pass filter, 1710–1880 MHz.

8

Coupling of Several Transceivers on the Same Antenna

8.1 Benefits of this Coupling

Chapter 4 discussed the coupling between antennas; it has shown us that to isolate, on a site, an antenna with vertical polarization from its neighbor it is required, between the two antennas, a minimum distance of

- λ in vertical.
- 3λ in horizontal.

To limit the height of the pylons or the clutter of the terraces, we will therefore have interest in connecting several transceivers on the same antenna.

We have already seen in Chapter 6 that solutions for the cellular systems with multiband panels that comprise in reality several antennas in the same radome, with one antenna and one access by frequency band, exist. Let's discuss the solutions for the coupling of several transceivers in the same frequency band on the same antenna.

This is the function of the multicoupler. The multicoupler is essential also in the cellular networks that use a lot of channels in the same frequency band. The advantage of the multicoupler is not limited to reducing the number of antennas on a pylon.

It also allows:

- The reduction of the intermodulation products (see Chapter 9) in the input stages of receivers and in the power amplifiers of the transmitters.
- One to obtain better isolation of the transmitters.
- One to filter out spurious frequencies created on the site or induced.

8.2 Different Types of Multicouplers

There are two types of multicouplers:

- The transmission multicoupler, which is also called a combiner.
- The reception multicoupler.

These two types of multicouplers are usually associated with antenna sites because the coupling involves almost always transceivers. These two multicouplers are then connected to the antenna through the duplexer.

8.3 Transmission Multicoupler

There exist various types of transmission multicouplers that are selected according to the coupling that needs to be solved.

The components of the transmission multicouplers are:

- The 3 dB couplers;
- The cavity couplers;
- The active couplers;
- The circulators.

8.3.1 3 dB Couplers

The 3 dB coupler allows for the coupling of two transmission paths to the same output.

Different types of couplers can be used. These are the most common equipment:

- The ring coupler.
- The $\lambda/4$ microstrip-line coupler.

8.3.1.1 Ring Coupler

A ring coupler is made with three $\lambda/4$ lines and one $3\lambda/4$ line mounted in ring (Figure 8.1).

A wave (current, voltage; see Appendix G) entering path (1) is divided into two waves running on the ring in the opposite direction.

On the output path (2), the difference of the paths traveled by these waves is $5\lambda/4 - \lambda/4 = \lambda$.

They recombine thus in phase and with a phase shift of $-\pi/2$ relative to the path (1).

On the output path (3), the difference of the paths traveled by the two waves is $4\lambda/4 - 2\lambda/4 = \lambda/2$

They are in phase opposition, therefore cancel each other.

On the output path (4), the two waves that have traveled each half a ring are in phase with a phase shift of $-3\pi/2$ relative to the path (1).

Thus an incoming wave in path (1) gives outgoing waves of equal amplitude on the paths (2) and (4).

In order not to mismatch the coupler, all ports must have an impedance of under 50 Ω (load or source impedance).

In particular, the output path (3) must always be loaded with a 50 Ω load.

In terms of energy, as the lines are lossless, the power supplied to the input path (1) is to be found by half (-3 dB) on the output paths (2) and (4).

We would demonstrate in the same way that an incoming wave in path (3) would be divided by half power in the paths (2) and (4).

So, if a transmitter is connected in path (1) and another in path (3), we may obtain on one of the two output paths (2) or (4) a mixture of signals from the two transmitters but with a half-power.

The unused output will have to be loaded by 50 Ω (power greater than or equal to half the sum of powers of the two transmitters).

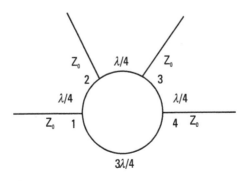

Figure 8.1   Ring coupler.

An example of an industrial ring coupler is given at the end of this chapter.

Exercise

Calculate the characteristic impedance Z_c of the lines $\lambda/4$ and $3\,\lambda/4$ to have the adaptation to access paths (1), (2), (3) and (4), knowing that Z_o is the characteristic impedance of the access lines.

Solution

The input impedance Z_e of a line $\lambda/4$ or $(2k+1)\,\lambda/4$, of characteristic impedance Z_0, loaded by a real impedance Z_R, is (see Appendix G).

$$Z_e = Z_c^2/Z_R$$

The matching condition to accesses is

$$Z_e = 2Z_0$$

Moreover,

$$Z_R = Z_0 Z_e/(Z_0 + Z_e) = 2Z_c^2/3Z_0 = 2/3Z_0$$

Consequently,

$$Z_c = \sqrt{Z_e Z_R} = 2Z_0/\sqrt{3}$$

8.3.1.2 $\lambda/4$ Microstrip-Line Coupler

The $\lambda/4$ microstrip line coupler is represented in Figure 8.2.

It consists of two $\lambda/4$ microstrip lines (see Appendix G) with their two conductive strips on top of the substrate strongly coupled (see Figure 8.3).

The characteristic impedance of each line is 70 Ω, but because of the coupling, the real characteristic impedance is 50 Ω.

It has two inputs (one for each transmitter) and two outputs: one toward the antenna and the other for a 50-Ohm load.

According to the theory of the *directional coupler* (see Appendix G), a wave injected in T_{x1} is going with half-power in the direction of the antenna and with half-power in the other microstrip line strongly coupled, but in the opposite direction; that is to say, in the load direction.

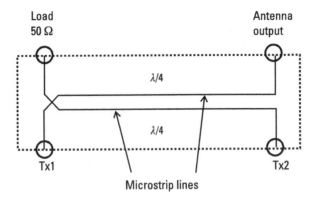

Figure 8.2 $\lambda/4$ microstrip-line coupler.

For the same reason, the wave injected in T_{x2} is going with half-power to the load and with half-power to the antenna.

Therefore, the half-power of each transmitter is provided to the antenna.

The power of the load must be greater than or equal to half the sum of the powers of the two transmitters.

This coupler is the most used on the mobile radio networks.

An industrial example is given at the end of this chapter.

8.3.2 Cavity Couplers

We have seen that a 3 dB coupler can couple two transmitters on the same antenna, but there is a loss of 3 dB on the power of each transmitter.

This power loss would be even higher (6 dB) if four transmitters on one antenna were coupled with 3 dB couplers (Figure 8.4).

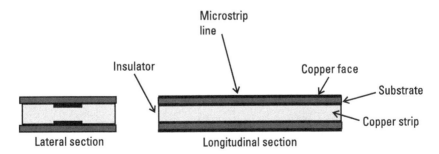

Figure 8.3 Microstrip lines coupling.

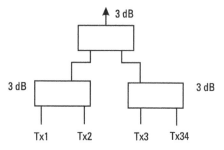

Figure 8.4 Coupling four transmitters on one antenna with 3 dB couplers.

This is why cavity couplers are very often used when the number of transmitters to be coupled is important.

These couplers are made up of coaxial cavities connected in a star (Figure 8.5).

These couplers have the advantage of having very low losses. This is because the insertion losses of coaxial cavities are of about 0.5 to 1 dB maximum, but they require minimum frequency differences between channels.

In 400 MHz, the spacing of the channels must be at least 200 kHz.

8.3.3 Active Couplers

These couplers are used to couple on a single antenna a large number of transmitters (more than 4).

They consist of cavities as previously with amplifiers in each branch (Figure 8.6).

The gain of the amplifiers is calculated such that the insertion loss is 0 dB on each path.

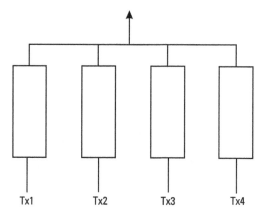

Figure 8.5 Cavities coupler.

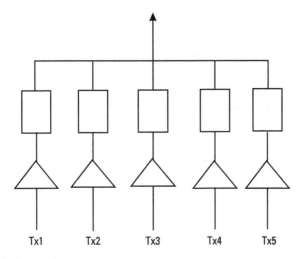

Figure 8.6 Active coupler.

8.3.4 The Circulators

The circulators are hexapoles and are widely used in the transmitter multicouplers for their property of nonreciprocity due to the use of ferrites (Figure 8.7).

An incoming wave in (1) will come out in (2), an incoming wave in (2) will come out in (3), and so on.

This is due to a property of the ferrite.

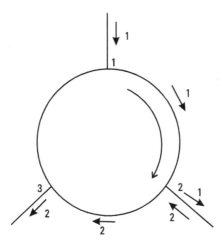

Figure 8.7 Principle of the circulator.

8.3.4.1 Property of the Ferrite.

The ferrite is an iron oxide into which an impurity is added. This impurity can be atoms of manganese, magnesium, or nickel.

The ferrite has a very high resistivity (from 10^6 to 10^{10} $\Omega \times$ cm) and magnetic permeability of several dozens of units.

That is why they have the applications that are already known: inductors, transformers in the circuits LF or HF (up to 30 MHz).

The nonreciprocity is another ferrite property used in the circulators.

For example, if we place a piece of ferrite in a waveguide in TE_{10} mode (see Appendix H) and if this ferrite is subjected to a continuous magnetic field H_0 parallel to the electric field and of well-defined value, an electromagnetic wave will pass in a direction of the guide and will be absorbed in the other direction.

This is called nonreciprocity.

This is the principle of the isolator used in hyper frequencies (SHF and EHF bands).

8.3.4.2 Achieving a Circulator

The circulator is, in general, of triplate technology. It is represented in Figure 8.8.

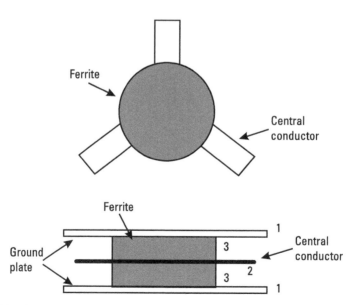

Figure 8.8   Achieving a circulator.

- The 2 parts 1 are the ground plates of the triplate.
- Part 2 is the central conductor connected to the 3 paths.
- The parts 3 are the ferrite cylinders placed between 1 and 2.

The continuous magnetic field H_0 is provided by permanent magnets that are not represented in the figure. The insertion loss in the forward direction is < 1 dB. The isolation between paths in the prohibited direction is 20–30 dB.

An industrial example of the circulator is given at the end of the chapter.

8.4 Receiving Multicoupler

The receiving multicoupler is an active multicoupler (Figure 8.9). It consists of:

- A band-pass filter, called a *preselector*;
- A very linear amplifier;
- As many power dividers as is necessary to obtain the number of desired paths.

The wide-band preselector filter protects the amplifier against the strong signals present on the site. This wide-band preselector filter is often insufficient in the case of a site loaded with other transmitters, and requires the addition of an extra narrower band-pass filter. The amplifier must be very linear so as not to create intermodulation products of third order (see Chapter 9). It compensates

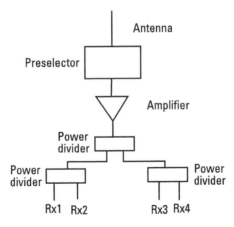

Figure 8.9 Receiving multicoupler.

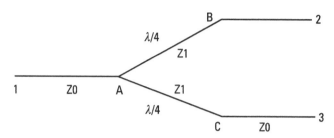

Figure 8.10   Power divider.

for the low loss of power dividers such that each path has an insertion loss of 0 dB. It must also have a low noise factor (2 dB maximum) in order to avoid damaging the signal-to-noise ratio. The power dividers are hexapoles made with two $\lambda/4$ lines AB and AC (Figure 8.10).

When a wave arrives through the path (1), it is divided between two equal parts in the paths (2) and (3), since the structure is symmetrical with respect to the path (1). In order to have each line Z_0 matched, the characteristic impedance Z_1 of the $\lambda/4$ lines must be such that their impedance in A is $2Z_0$. According to the formula in Appendix G (G.16).

$$\bar{Z}_x = \bar{Z}_0(\bar{Z}_R + Z_0 - \bar{Z}_R + Z_0)/(\bar{Z}_R + Z_0 + \bar{Z}_R - Z_0) = Z_0^2/\bar{Z}_R$$

$$\bar{Z}_x = Z_0^2/\bar{Z}_R$$

with $Z_x = Z_0$ (impedance in B and C) and $Z_0 = Z_1$ (characteristic impedance of $\lambda/4$ lines).

Therefore, $Z_1 = \sqrt{2}Z_0$.

8.5 The Duplexer

The duplexer is used to separate the transmission path and the reception path at the antenna level. There are three classes of duplexers:

- The notch filter duplexers;
- The band-stop filter duplexers;
- The band-pass filter duplexers.

8.5.1 Notch Filter Duplexer

The notch filter duplexer is made with two notch filters connected to the antenna (Figures 8.11 and 8.12).

The spacing between the transmission frequency and reception frequency (duplex spacing) is 10 MHz in the 400 MHz band. The role of the notch filter on the transmission path is to let the transmission frequency pass and to reduce the reception frequency by at least 60 dB.

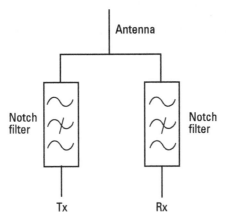

Figure 8.11 Notch filter duplexer.

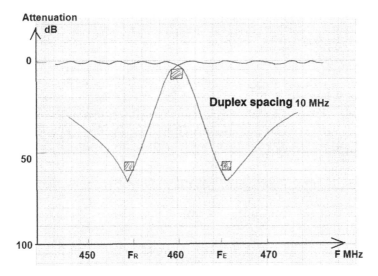

Figure 8.12 Response curve of the duplexer.

The role of the notch filter on the reception path is to let the reception frequency pass and to reduce the transmission frequency by at least 60 dB. These filters are manufactured in the VHF bands 34–41 MHz and 68–88 MHz with LC components, and in the VHF band 151–162 MHz and UHF band (300–3000 MHz) with coaxial cavities (see Chapter 7). An example of notch duplexer is given at the end of this chapter.

8.5.2 Stop-Band Filter Duplexer

There are also stop-band filter duplexers which, instead of simply rejecting a single frequency in each path, reject a small band of frequencies. These duplexers are used in the multiplexers with several transceivers (see Figures 8.15 and 8.16 in Section 8.6).

8.5.3 Band-Pass Filter Duplexer

The notch filter duplexers and the stop-band filter duplexers have the disadvantage of rejecting only one frequency (or, in some cases, a small band of frequencies) in each direction, and letting all the others pass. For this reason, the duplexers using these filters are not recommended for antenna sites heavily loaded in transmitters-receivers, or located in an area where there are many spurious frequencies.

It is necessary, in that case, to use band-pass filter duplexers. The duplexers shown in Figures 8.13 and 8.14 are made with two band-pass filters connected to the antenna.

The role of the band-pass filter on the transmission path is to let the transmission frequency and a limited band around this frequency pass, and to reduce the reception frequency by at least 60 dB (in this example, 65 dB minimum). The role of the band-pass filter on the reception path is to let the reception frequency and a limited band around this frequency pass, and to reduce the transmission frequency by at least 60 dB (in this example, 65 dB minimum).

These duplexers are used in two cases:

- When the transmitter-receiver is installed on a site loaded with antennas.
- On the transmission-reception multicouplers. This is because several frequencies must be passed on the two paths.

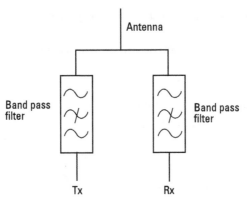

Figure 8.13 Band-pass filter duplexer.

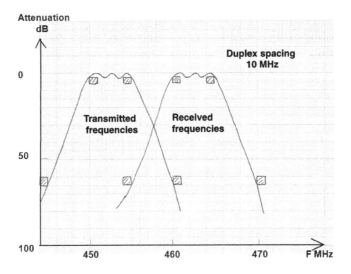

Figure 8.14 Response curve of the band-pass duplexer.

These filters are manufactured in the VHF 34–41 MHz and 68–88 MHz bands with LC components and in the VHF 151–162 MHz and UHF bands with coaxial cavities (see Chapter 7). An example of a band-pass duplexer is given at the end of this chapter.

8.6   Achieving a Transceiver Multicoupler

These multicouplers can use the following two technologies for the transmission part:

- The 3 dB coupler technology.
- The cavity coupler technology.

8.6.1   3 dB Coupler Technology

The schematic diagram of a multicoupler for four transceivers is represented in Figure 8.15.

We can note that:

- Three 3 dB couplers are used. These 3 dB couplers are with four inputs/outputs. The output that is not used must be loaded by a 50 Ω load of power equal to half of the total power of the two transmitters connected to the coupler.
- Two circulators by transmission path provide a minimum isolation of 40 dB between two transmitters connected to the same 3 dB coupler.
- The 50 Ω loads of the circulators must be able to withstand the power of a transmitter.
- The duplexer must be with band-pass filters.
- The receiver multicoupler is a standard four-path multicoupler.
- The insertion loss of each transmission path is 7 dB.

8.6.2   Cavity Coupler Technology

The schematic diagram of a multicoupler for four transceivers is represented in Figure 8.16.

In this case, the main difference with the other type of technology is the low-insertion loss in each path: 2 dB. That is why this technology is particularly recommended for multicouplers with more than four paths.

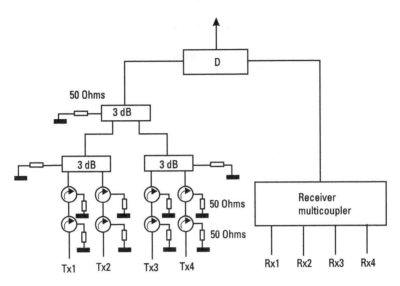

Figure 8.15 3 dB coupler technology.

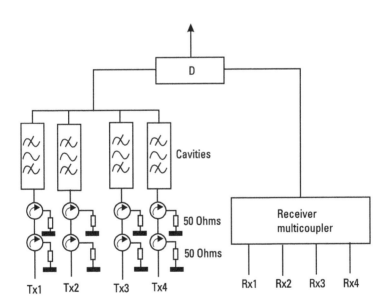

Figure 8.16 Cavity coupler technology.

Reference

[1]   Piette, B., *Multicouplers and Filters VHF/UHF*, France: Lavoisier, 2007.

Appendix 8A

Examples of Industrial Equipment

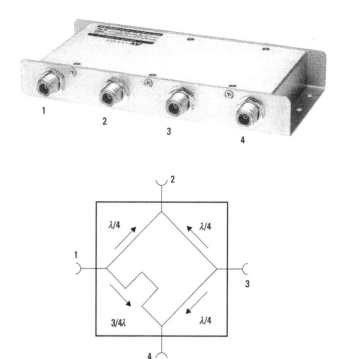

Figure 8.17   Ring coupler, 400–470 MHz.

Technical Characteristics

- Frequency range: 400–470 MHz.
- Attenuation: 1/2, 1/4: 3+/−0.4 dB.
- Attenuation: 3/2, 3/4: 3+/− 0.4 dB.
- Isolation: 1/3, 2/4: 30 dB.
- VSWR < 1.2.
- Impedance: 50 Ω.
- Input power < 100W.
- Dimensions: 225 mm × 32 mm × 117 mm.

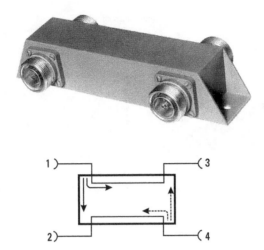

Figure 8.18 3 dB microstrip coupler, 340–512 MHz.

Technical Characteristics

- Frequency range: 340–512 MHz.
- Attenuation 1/2, 1/3: 3+/−0,4 dB.
- Attenuation 4/2, 4/3: 3+/−0,4 dB.
- Isolation 2/3, 1/4 : 30 dB.
- VSWR < 1.06.
- Impedance: 50 Ω.
- Input power < 500W.
- Dimensions: 252 mm × 47 mm × 115 mm.

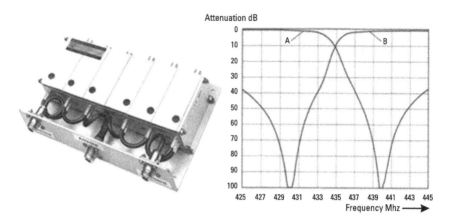

Figure 8.19   Notch duplexer, 380–470 MHz.

Technical Characteristics

- Number of coaxial cavities: 3 + 3.
- Frequency band: 380–470 MHz.
- Duplex spacing: 10 MHz.
- Insertion loss < 0.7 dB.
- VSWR < 1.4.
- Impedance: 50 Ω.
- Input power < 50W.
- Dimensions 270 mm × 58 mm × 190 mm.

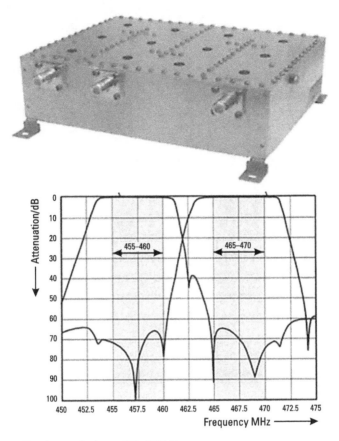

Figure 8.20 Band-pass duplexer, 455–470 MHz.

Technical Characteristics

- Number of coaxial cavities: 6 + 6.
- Frequency band: 455–460 MHz and 465–470 MHz.
- Duplex spacing: 10 MHz.
- Insertion loss < 0.8 dB.
- VSWR < 1.25.
- Impedance: 50 Ω.
- Input power < 200W.
- Dimensions: 315 mm × 87.5 mm × 244.6 mm.

9

Study of the Radio Environment on an Antenna Site, Measurements of the Spurious Frequencies, and Protection against These Frequencies

9.1 The Purpose of this Chapter

Nowadays, mobile radio networks use duplex infrastructures consisting of the number of antenna sites necessary to achieve the desired coverage area and for cellular networks, private (TETRA, TETRAPOL, APCO 25, etc) and operated (GSM, IS95, IS136, PDC, UMTS, CDMA 2000, etc), to ensure the required traffic with a good quality of service. The development of mobile telephones over the last twenty years has led to a large deployment of antenna sites in the world and, inevitably, to an increasing number of antennas on each site. It follows that there are interference frequencies on an antenna site that may debase significantly the performance of the receivers that we would like to install. These interference frequencies are:

- The frequencies of the other transmitters in the site;
- The noise of these transmitters;
- The harmonics of these transmitters;
- The intermodulation products of third order created by the power of these transmitters;
- The spurious frequencies from outside the site.

The higher the number of transmitters-receivers located on the site correlates to the greater the risks to see spurious frequencies. Fortunately, there are means to overcome these problems. This is why a new installation of an antenna on a site requires a study of the radio environment of the transmitter-receiver that is going to be installed, and of the solutions to solve the problems related to this environment.

9.2 Recommended Method for this Study

The recommended method for this study is divided into three phases: the information phase, the calculation phase, and the phase of measurements and research of protection solutions.

9.2.1 Information Phase

Pieces of information about the following topics are to be taken:

- The companies that have installed the radio networks on the site;
- The frequencies of the transmitters already installed on the site;
- The power of these transmitters.

9.2.2 Calculation Phase

We calculate the harmonic 2 and the harmonic 3 of each of the transmitter frequencies. Using specialized software, we calculate all intermodulation products of third order that can be created by these frequencies.

9.2.3 Phase of Measurements and Research of Protection Solutions

Measurements on the transmitter:

- For measurement of the antenna VSWR, see Appendix G.

Measurements in the receiver input:

- Measurement of the powers received from the transmitters;
- Measurement of the transmitters' noise;

- Measurement of the transmitters' harmonics;
- Measurement of the intermodulation products of third order;
- Measurement of the frequencies from outside the site.

9.3 Measurement on the Transmitter—Antenna VSWR

This measurement consists of controlling the installation of the antenna. It is done with a VSWR-meter that measures the forward power and the reflected power (see Appendix G) using an integrated directional coupler, which deduces the VSWR displayed on a digital screen. The VSWR is measured by inserting the apparatus between the transceiver output and the coaxial cable that goes to the antenna. It must be less than 1.5 when the transmitter is transmitting. If the measurement exceeds 1.5, it is necessary to check the connection of the antenna to the coaxial cable and check the cable's connectors.

9.4 Measurement of the Spurious Frequencies on the Receiver and Protection against these Frequencies

9.4.1 Setups of the Measurements to be Carried Out

Two measurement patterns must be carried out, depending on whether the signals to be measured are strong or weak.

9.4.1.1 Strong Signals

To measure the power of the transmitters of the site at the receiver input, the pattern n°1 will be used in Figure 9.1.

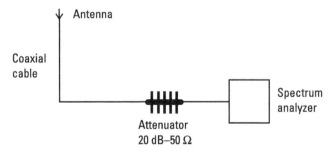

Figure 9.1   Setup of measurement n°1—strong signals.

The coaxial cable of the antenna base station or repeater is disconnected from the transceiver, and connected to the analyzer through an attenuator 20 dB.

9.4.1.2 Weak Signals

The weak signals to be measured are:

- The noise of the transmitters installed on the site;
- The harmonics of the site transmitters;
- The intermodulation products of third order generated by the transmitters interfering with each other;
- The spurious frequencies from outside the site.

The pattern used for the measurement of these weak signals is the pattern n°2 shown in Figure 9.2.

A 10 dB attenuator and a narrower band-pass cavity filter make it possible to limit the strong signals at the input of the analyzer. One or more notch filters may be required, in addition to the band-pass filter, in order to avoid the saturation of the HF head of the analyzer.

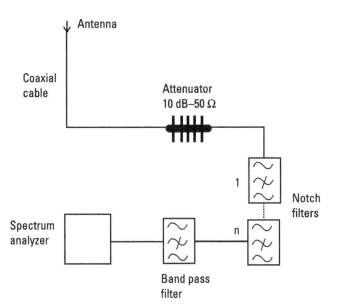

Figure 9.2   Setup of measurement n°2—weak signals.

9.5 Measurement of the Frequency Levels of the Site Transmitters and Protection against Blocking

9.5.1 Measurement

We use the setup n°1 and we measure the level of these frequencies on the analyzer. We add the 20 dB of the attenuator, and we check that the levels are below the level of *blocking* (saturation of the receiver) given by the specifications: 90 dB over the sensitivity of the receiver.

9.5.2 Protection

If one or more levels are superior to the limit, one or more notch filters will have to be added at the receiver input.

9.6 Measurement of the Noise of the Transmitters and Protection against this Noise

The noise of the transmitters is generated mainly by:

- The harmonic distortion caused by the lack of linearity or by the saturation of the transmitter modulators. We then obtain spectral lines on an analyzer.

- The phase and amplitude noise of the oscillators, particularly with the frequency synthetizers. We then obtain white noise on an analyzer.

This noise is mostly troublesome when there are very powerful transmitters, such as broadcast FM transmitters, on the site. This noise can also be troublesome when the installation includes a receiver in the 80 or 150 MHz bands. For the transmitters of the mobile radio networks, the duplex spacing is important: 4.6 MHz in the bands 40 MHz, 80 MHz, 150 MHz; 10 MHz in the band 400 MHz; 45 MHz for GSM 900 MHz; 95 MHz for GSM 1800 MHz; and so on. The reception frequencies are far from the transmission frequencies. Therefore, the transmitter noise should not be a problem. However, this must be verified all the same.

9.6.1 Measurement

The measurement setup used is the setup n°2, wherein it is necessary to introduce all the required notch filters in order to remove the strong signals.

According to whether we consider spectral lines or white noise, we use one of the two following methods:

1. *The First Method:* If they are spectral lines, the level of the lines in the bandwidth of the receiver is measured on the analyzer. The sum of the powers of each line is calculated in dBm. The level of this sum must not exceed the limit specified in the specifications for the protection of the useful path, namely:

   - *For an analog receiver in 25 KHz:* 8 dB below the value of the sensitivity of the receiver;

   - *For an analog receiver in 12.5 KHz:* 11 dB below the value of the sensitivity of the receiver;

   - *For a TETRA digital receiver in 25 KHz:* 19 dB below the value of the sensitivity of the receiver;

   - *For a TETRAPOL digital receiver in 12.5 KHz:* 15 dB below the value of the sensitivity of the receiver;

   - *For a GSM digital receiver in 200 KHz:* 15 dB below the value of the sensitivity of the receiver.

2. *The Second Method:* If it is white noise, then it is necessary to begin by getting rid of the noise of the spectrum analyzer. For that, this measuring device must be set to a bandwidth of 500 Hz, for example, so that the noise floor of the device is at least 6 dB below the noise to be measured. We then measure that level, which—on some spectrum analyzers—can be measured directly as power per Hertz. If the analyzer does not allow it, we measure the noise power P_{500} Hz in the band 500 Hz, and we bring it back to power per Hertz by

$$P_{\mathrm{Hz}} \text{ (dBm)} = P_{500 \text{ Hz}} \text{ (dBm)} - 10 \log 500 = P_{500 \text{ Hz}} \text{ (dBm)} - 27 \text{ dB}$$

We then have to calculate the power in the bandwidth BP of the receiver by

$$P_{\mathrm{BP}} \text{ (dBm)} = P_{\mathrm{Hz}} \text{ (dBm)} + 10 \log \text{ BP (Hertz)}$$

P_{BP} must be below the limit of protection on the useful path indicated above.

9.6.2 Protection

In case the limit is exceeded, the only means of protection would be to move the antenna. It would also be possible to negotiate with the company that installed the transmitter in question for a resumption of the settings of this one.

9.7 Measurement of the Harmonics of the Transmitters and Protection against these Spurious

9.7.1 Measurement

Setup n°2 is used with the notch filters that are necessary to remove the strong signals. Knowing the frequencies of all the transmitters of the site, we measure the level of the harmonics present in the bandwidth of the receiver. The sum of the powers of these harmonics must not exceed the level of specifications for the protection on the useful path, as indicated in Section 9.6.1.

9.7.2 Protection

In case the limit is exceeded, the only means of protection would be to move the antenna or to negotiate with the company (or companies) that installed the transmitters for the addition of notch filters on the antenna output of the latter.

9.8 Measurement of the Intermodulation Products of Third Order and Protection

The intermodulation products of third order result from the nonlinearity of the power amplifiers of the transmitters and of the input stages of the receivers. They can be created by the frequencies of the transmitters installed on the site, either in the transmitter itself, or in the receiver. If ΔF is the difference between the frequencies of the transmitters F_1 and F_2, the intermodulation products of third order are $2F_1 - F_2$ and $2F_2 - F_1$. They are located on either side of F_1 and F_2 at the distance ΔF (see Figure 9.3).

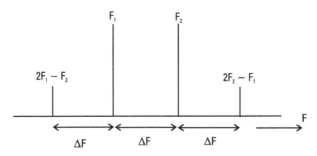

Figure 9.3 Intermodulation products of third order.

9.8.1 Intermodulation Products of Third Order in the Transmitter

9.8.1.1 Measurement

Setup n°2 is used with the notch filters that are necessary to remove the strong signals. From the knowledge of the transmitter frequencies, all of the intermodulation products of third order possible on the site were calculated using specialized software. We identify those that might fall within the receiver bandwidth and we search for them on the analyzer. We verify that the sum in dBm of those present has a level lower than that of the specifications for protection on the useful path (this is discussed in Section 9.6.1).

9.8.1.2 Protection

In case the limit is exceeded, the only means of protection would be to move the antenna, or to negotiate with the company (or companies) that which installed the transmitters for the addition of notch filters or circulators (see Chapter 8) to the output of the latter.

9.8.2 Intermodulation Products of 3rd Order in the Receiver

9.8.2.1 Measurement

Measurement setup n°1 is used. We read on the analyzer the levels in dBm of the two frequencies F1 and F2, which give a product in the band of the receiver, and we add 20 dB to the results (attenuator). The specifications require a protection of 70 dB above the receiver sensitivity, for two signals F1 and F2 at the same level. It is unlikely that the two levels are equal in our measurement, however we are certain that, if no level is above 70 dB with respect to the receiver sensitivity, the receiver is protected.

9.8.2.2 Protection

In case the limit is exceeded on a single frequency, it would be necessary to add a notch filter on that frequency to the receiver input. In case the limit is exceeded on the two frequencies, it would be necessary to add a notch filter on one of the two frequencies.

9.9 Measurement of the Spurious Frequencies from Outside the Site and Protection

9.9.1 Measurement

We use measurement setup n°2 and we examine the analyzer in the receiver bandwidth to see if there are other spectral lines. If there are spectral lines, then the sum in dBm of those lines must be at a level lower than the level of the specifications for protection on the useful path, which is discussed in Section 9.6.1.

9.9.2 Protection

If the limit is exceeded, there is no means of protection, and a report must be made to the Office of Frequencies Management.

10

Radiating Apertures Horn and Parabolic Antennas

1.1 Radiating Apertures

The radiation of plane apertures finds its application in different types of antennas; in particular, horns and parabolic antennas.

Parabolic antennas, which use a horn as an excitation source, are widespread in mobile phone infrastructures. Indeed, very often, the antenna sites are connected to the network switch by microwave links (not by wire connections) because it is more expensive to have telephone lines in an antenna site, especially in rural areas. The parabolic antennas are powered by waveguides (see Appendix H).

In the past, a waveguide was installed up the pylon between the transceiver of the microwave link terminal and the parabolic antenna. Since the removal of electronic tubes in the power amplifiers of the transmitters and their replacement by semiconductors, these amplifiers are much smaller and lighter. They can be installed with the receiver front end in a housing at the back of the parabolic antenna. In these conditions, a coaxial cable that conveys two frequencies in the range of 100 to 200 MHz (one for transmission and the other for reception) is used between the transceiver and this housing. A frequency change must be made in the housing to convert these frequencies to the transmission and reception frequencies. Thus, the waveguide is now only used between the housing and the antenna. This waveguide is connected to a horn, placed at the focus of the paraboloid, and provides a spherical wave to the antenna, which the latter turns into a plane wave (see Chapter 1 for the definition of the different types of waves).

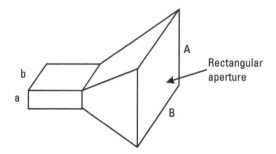

Figure 10.1 Pyramidal horn.

Three types of horns are used in the parabolic antennas:

1. The sectorial horn;
2. The pyramidal horn;
3. The conical horn.

Here we will present only the pyramidal horn (Figure 10.1), which is used most often.

1.2 Huyghens-Fresnel Formula

Let us present any plane aperture of surface S in the xoy plane (Figure 10.2). We illuminate this aperture by a wave beam coming from the region of the negative z.

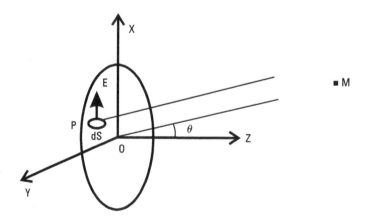

Figure 10.2 Radiating aperture.

There exist at each point P of (S):

- An electric field $\vec{E}(x,y)$;
- A magnetic field $\vec{H}(x,y)$.

We will assume:

- The aperture of large dimension regarding λ;
- That all the vectors $\vec{E}(x,y)$ are polarized vertically (parallel to ox);
- That point M is away from the aperture (r greater than $10\ \lambda$) and the rays are slightly inclined to the axis oz (θ small).

According to the principle of Huyghens-Fresnel:

- Each point P of the aperture (S) acts as a secondary source whose phase is that of the incident wave at the same place.
- The field at point M will be the vector sum of the fields produced by all the points P of the aperture.
- The elementary field at M, produced by an element dS surrounding the point P, is calculated by Fresnel's formula:

$$\overline{dE} = jE(x,y)/\lambda r e^{-j\beta r} e^{j\beta OP\cos(\text{angle } OP,OM)} dS$$

with $B = 2\pi/\lambda$.

The field at a point M in space will therefore be the vector sum of the fields from all the points P of (S).

Let us make the following calculation assumptions (Figure 10.3):

- $\vec{i}, \vec{j}, \vec{k}, \vec{u}$ are the unit vectors of ox, oy, oz, OM;
- Φ, Ψ, θ are the OM angles with the axes ox, oy, oz.

We can write, according to the definition of the scalar product (see Appendix B):

$$OP\cos(\text{angle } OP,OM) = \overrightarrow{OP}.\vec{u} = x\vec{i}.\vec{u} + y\vec{j}.\vec{u} = x\cos\Phi + y\cos\Psi$$

The total field provided at M created by the aperture is thus:

$$\bar{E} = j/\lambda r e^{-j\beta r} \iint_S E(x,y)e^{j\beta(x\cos\Phi + y\cos\Psi)} dS \tag{10.1}$$

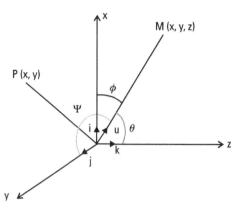

Figure 10.3 Calculation assumptions.

We will therefore apply this formula to the two radiating apertures:

- The rectangular aperture (application to the pyramidal horn);
- The circular aperture (application to the parabolic antenna).

1.3 Radiating Rectangular Aperture: Application to Pyramidal Horn

We will assume:

- That the rectangular aperture is that of a pyramidal horn and that its surface is AB (Figure 10.1).
- That the horn is connected by a rectangular waveguide in TE_{10} mode (see Appendix H) of section ab.
- That the field at each point of the horn aperture follows the same law in amplitude and in phase as on the output surface ab of the waveguide.

1.3.1 Radiation Patterns

1.3.1.1 Pattern in the E-Plane

We have represented the rectangular aperture of the horn in the xoy-plane (Figure 10.4), as well as any point P on the aperture with its electric field $\vec{E}$ and a remote point M in the xoz-plane. The aperture is polarized according to ox (electric fields parallel to ox). The E-plane is the xoz-plane. In this plane $\Psi = \pi/2$ and we can replace Φ by $\pi/2 - \theta$.

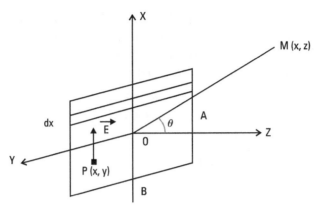

Figure 10.4   Rectangular aperture.

Equation (10.1) thus becomes:

$$\bar{E} = j/\lambda r \; e^{-j\beta r} \iint_S E(x, y) e^{j\beta x \sin\theta} \; dS$$

The double integral I can be transformed into two simple integrals by successively integrating with x = constant and with y = constant.

$$I = \int_{-B/2}^{+B/2} E(x, y) \, dy \; \int_{-A/2}^{+A/2} e^{j\beta x \sin\theta} \, dx$$

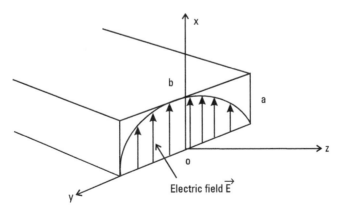

Figure 10.5   Waveguide TE_{10} mode.

We have assumed that the field E (x, y) follows the same law in amplitude and phase as the electric field E in the fundamental mode TE_{10}, inside of a wave section ab (see (H.4) in Appendix H). However, the coordinates x and y are reversed in regard to the change of axis.

$$\vec{E} \left| \begin{array}{l} \bar{E}_x = 2jE_0 \cos(\pi y/b)e^{j\omega t}e^{-j2\pi z/\lambda g} \\ E_y = 0 \\ E_z = 0 \end{array} \right. \qquad (10.2)$$

We would therefore have the same coordinates on the aperture by changing b into B.

We can write:

$$\int_{-B/2}^{+B/2} E(x,y)\, dy = 2E_0 \int_{-B/2}^{+B/2} \cos(\pi y/B)\, dy = 4BE_0/\pi$$

What remains is to calculate the second integral: $\int_{-A/2}^{+A/2} e^{j\beta x \sin\theta}\, dx$
We find:

$$\left(e^{+j\beta \frac{A}{2} \sin\theta} - e^{-j\beta \frac{A}{2} \sin\theta} \right) \bigg/ j\beta \sin\theta$$

$$= 2\sin\left(\frac{\beta A}{2}\sin\theta\right) \bigg/ \frac{\beta A}{2}\sin\theta$$

because $\sin\alpha = (e^{j\alpha} - e^{-j\alpha})/2j$.
We finally obtain here the modulus of $\vec{E}$ by replacing β by $2\pi/\lambda$:

$$E = 4ABE_0/\pi\lambda r \left[\sin(\pi A\sin\theta/\lambda)/(\pi A\sin\theta/\lambda)\right] \qquad (10.3)$$

We can note that the electric field in the E-plane depends only on A. The characteristic function, enabling to plot the radiation pattern in the E-plane is thus:

$$F(\theta) = \sin(\pi A\sin\theta/\lambda)/(\pi A\sin\theta/\lambda) \qquad (10.4)$$

which is of the form sin u/u.

To draw this pattern in polar coordinates, we will use the table giving sin x/x as a function of x (Table 10.1) with $x = \pi A \sin\theta/\lambda$. The pattern will be presented in an example in Section 10.5.

1.3.1.2 Pattern in the H Plane

The H plane by definition (see Chapter 1) is the plane formed by the magnetic field and the propagation direction oz. It is the yoz-plane.

In this plane $\Phi = \pi/2$ and we can replace Ψ by $\pi/2 - \theta$.

Table 10.1
Table sin x/x

x	sin(x)/x	x	sin(x)/x	x	sin(x)/x
0.0	1.00000	2.8	0.11964	5.6	−0.11273
0.1	0.99833	2.9	0.08250	5.7	−0.09661
0.2	0.99335	3.0	0.04704	5.8	−0.08010
0.3	0.98507	3.1	0.01341	5.9	−0.06337
0.4	0.97355	3.2	−0.01824	6.0	−0.04657
0.5	0.95885	3.3	−0.04780	6.1	−0.02986
0.6	0.94107	3.4	−0.07516	6.2	−0.01340
0.7	0.92031	3.5	−0.10022	6.3	0.00267
0.8	0.89670	3.6	−0.12292	6.4	0.01821
0.9	0.87036	3.7	−0.14320	6.5	0.03309
1.0	0.84147	3.8	−0.16101	6.6	0.04720
1.1	0.81019	3.9	−0.17635	6.7	0.06042
1.2	0.77670	4.0	−0.18920	6.8	0.07266
1.3	0.74120	4.1	−0.19958	6.9	0.08383
1.4	0.70389	4.2	−0.20752	7.0	0.09385
1.5	0.66500	4.3	−0.21306	7.1	0.10267
1.6	0.62473	4.4	−0.21627	7.2	0.11023
1.7	0.58333	4.5	−0.21723	7.3	0.11650
1.8	0.54103	4.6	−0.21602	7.4	0.12145
1.9	0.49508	4.7	−0.21275	7.5	0.12507
2.0	0.45465	4.8	−0.20753	7.6	0.12736
2.1	0.41105	4.9	−0.20050	7.7	0.12833
2.2	0.36750	5.0	−0.19179	7.8	0.12802
2.3	0.32422	5.1	−0.18153	7.9	0.12645
2.4	0.28144	5.2	−0.16990	8.0	0.12367
2.5	0.12939	5.3	−0.15703	8.1	0.11974
2.6	0.19827	5.4	−0.14310	8.2	0.11472
2.7	0.15829	5.5	−0.12828	8.3	0.10870

Therefore, in this case, the Fresnel formula (10.1) becomes

$$\bar{E} = j/\lambda r \ e^{-j\beta r} \iint_S E(x, y) e^{j\beta y \sin\theta} \, dS$$

As seen in the previous section,

$$E(x, y) = 2E_0 \cos(\pi y/B) \qquad (10.5)$$

Let us replace $E(x, y)$ by its value in the double integral:

$$\bar{E} = 2jE_0/\lambda r \ e^{-j\beta r} \iint_S \cos(\pi y/B) e^{j\beta y \sin\theta} \, dS$$

The double integral I can be transformed into two simple integrals by integrating successively with $x = $ constant and $y = $ constant.

$$I = \int_{-A/2}^{+A/2} dx \ \int_{-B/2}^{+B/2} \cos \pi y/B \ e^{j\beta y \sin\theta} \, dy$$

$$I = A \int_{-B/2}^{+B/2} \cos \pi y/B \ e^{j\beta y \sin\theta} \, dy$$

We finally obtain here the modulus of $\vec{E}$ by replacing β by $2\pi/\lambda$:

$$E = 4ABE_0/\pi\lambda r \ \left[\frac{\pi^2}{4} \cos(\pi B \sin\theta/\lambda) \Big/ \left(\frac{\pi^2}{4} - \pi^2 B^2 \sin^2\theta/\lambda^2\right)\right] \qquad (10.6)$$

We also note that the electric field in the H-plane depends only on B. The characteristic function enabling to draw the radiation pattern in the H-plane is then:

$$F(\theta) = \frac{\pi^2}{4} \cos(\pi B \sin\theta/\lambda) \Big/ \left(\frac{\pi^2}{4} - \pi^2 B^2 \sin^2\theta/\lambda^2\right)$$

This formula is of the form:

$$F(\theta) = \frac{\pi^2}{4} \cos x \Big/ \left(\frac{\pi^2}{4} - x^2\right) \qquad (10.7)$$

by writing $x = \pi B \sin\theta/\lambda$.
Table 10.2 provides $F(\theta)$ as a function of x.

Table 10.2
$F(\theta)$ as a Function of x

x	F(θ)	x	F(θ)
0	1	2.8	0.43
0.4	0.99	3	0.38
0.8	0.94	3.2	0.32
1.2	0.86	3.4	0.24
1.6	0.8	3.6	0.21
1.7	0.76	3.8	0.20
1.8	0.7	4	0.18
2	0.66	4.2	0.08
2.2	0.61	4.3	0.06
2.4	0.54	4.5	0.03
2.6	0.49	4.7	0

The pattern will be represented on an example in Section 10.5.

1.3.2 Calculus of the Isotropic Gain of the Rectangular Aperture

The isotropic gain of the rectangular aperture in the main direction of propagation (*oz* axis) is given by the formula (see Chapter 1):

$$G_i = 4\pi U/P_e$$

where U is the radiation intensity in the oz direction and P_e is the power supplied to the aperture. U is given by the formula (see Chapter 1):

$$U = Wr^2$$

where W is the radiated power per unit area around the point M in the direction of the main propagation, and r is the distance OM.

In Appendix D, we find the formula $W = \varepsilon E^2/2$. For $\theta = 0$ (main propagation direction), we obtain, from (10.3) and (10.6)

$$E = 4ABE_0/\pi\lambda r$$

Hence

$$W = 8\varepsilon A^2 B^2 E_0^2 /\pi^2 \lambda^2 r^2$$

Therefore

$$U = 8\varepsilon A^2 B^2 E_0^2 / \pi^2 \lambda^2$$

What remains is to calculate the power P_e supplied to the aperture: The power provided by the waveguide to each surface element $dx\,dy$ of the horn aperture is

$$dW = 1/2\varepsilon E^2 dxdy$$

E being the electric field provided by the waveguide in each point of the horn aperture $E = 2E_0 \cos(\pi y/B)$ (see 10.5).
Hence

$$dW = 2\varepsilon E_0^2 \cos^2(\pi y/B)dxdy = \varepsilon E_0^2[1 + \cos(2\pi y/B)]dxdy$$

The power P_e is equal to

$$P_e = \varepsilon E_0^2 \int_{-A/2}^{+A/2} dx \int_{-B/2}^{+B/2}[1 + \cos(2\pi y/B)]\,dy$$

As the second integral is = 0, the total power supplied to the aperture is

$$P_e = \varepsilon ABE_0^2$$

Hence the isotropic gain of the rectangular aperture is

$$G_i = 4\pi U/P_e = 0.81(4\pi AB/\lambda^2).$$

1.4   Radiating Circular Aperture: Application to Parabolic Antenna

The output aperture of the parabolic antenna is an example of a radiating circular aperture. The parabolic antenna is illuminated by a spherical wave provided by a horn placed in the focus of the paraboloid. It provides a plane wave (see Chapter 1) thanks to a property of the paraboloid.

A paraboloid of vertex O, of focus F, of focal length OF $= d$ and of axis *oz* is represented in section in Figure 10.6.

A wave from F is reflected in M parallel to *oz*. Passing in P in the plane perpendicular to OF, the wave has traveled the distance FM + MP, which satisfies the relation:

$$FM + MP = 2d$$

The traveling of all the waves from F being equal, the Δ plane is equiphase and the waves from the paraboloid are plane waves. Therefore the radiating output aperture of the paraboloid provides a plane wave.

The feed of the parabolic antenna in F is a waveguide in TE_{10} mode terminated by a pyramidal horn. The latter provides spherical waves to the paraboloid.

1.4.1 Radiation Pattern of a Radiating Circular Aperture

The circular aperture being of revolution around the axis *oz*, the radiation patterns in the E- and H-planes are identical. We have represented (Figure 10.7) the radiating circular aperture in the *xoy* plane, as well as any point P on the aperture with its electric field $\vec{E}$ and a remote point M in the *xoz*-plane.

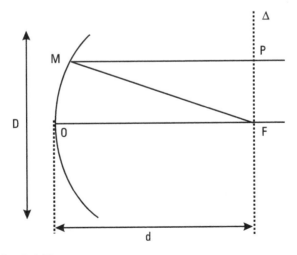

Figure 10.6 Paraboloid.

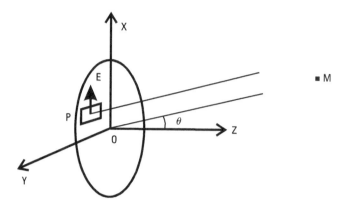

Figure 10.7   Radiating circular aperture.

Let $dx.dy$ be a small surface element surrounding the point P. The field it radiates in M is according to (10.1):

$$\bar{E} = j/\lambda r \; e^{-j\beta r} \iint_S E(x,y)e^{j\beta x \sin\theta}dS$$

since $\Psi = \pi/2$ and $\Phi = \pi/2 - \theta$.

$E(x,y)$ being uniform (same amplitude, same polarization, same phase) we can replace $E(x,y)$ by E_0:

$$\bar{E} = jE_0/\lambda r \; e^{-j\beta r} \iint_S e^{j\beta x \sin\theta}dS \qquad (10.8)$$

In this case the integration is first made with a constant x between N and N' (Figure 10.8). That is, between $y(N)$ and $y(N')$:

$$y(N) = \sqrt{a^2 - x^2}$$
$$y(N) = -\sqrt{a^2 - x^2}$$

This amounts to calculate dS. It then remains to calculate the integral for x varying from $-a$ to $+a$:

$$E = jE_0/\lambda r \; e^{-j\beta r} \int_{-\sqrt{a^2-x^2}}^{\sqrt{a^2-x^2}} dy \int_{-a}^{+a} e^{j\beta x \sin\theta} \, dx$$

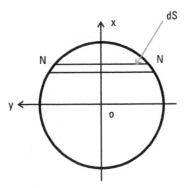

Figure 10.8   Integration.

that is to say

$$\bar{E} = 2jE_0/\lambda r\ e^{-j\beta r}\int_{-a}^{+a}\sqrt{a^2-x^2}\,e^{j\beta x\sin\theta}\,dx$$

Hence, by integrating and replacing β by $2\pi/\lambda$

$$\bar{E} = 2j\pi a^2 E_0/\lambda r\ e^{-j2\pi r/\lambda}\left[J_1(2\pi a\sin\theta/\lambda)/(2\pi a\sin\theta/\lambda)\right]$$

J_1 is a Bessel function.
Let us write $m = 2\pi a\sin\theta/\lambda$.
We finally get:

$$E = 2j\pi a^2 E_0/\lambda r\ e^{-j2\pi r/\lambda}J_1(m)/m \qquad (10.9)$$

$J_1(m)/m$ is given in the tables of the Bessel functions for all the values of m, replacing x by m (Table 10.3).
For $\theta = 0$, $J_1(m)/m = 1/2$, the field is maximum and its modulus is

$$E_{oz} = \pi a^2 E_0/\lambda r \qquad (10.10)$$

The characteristic function can be written as:

$$F(\theta) = E/E_{oz} = 2J_1(m)/m \qquad (10.11)$$

with $m = 2\pi a\sin\theta/\lambda$.

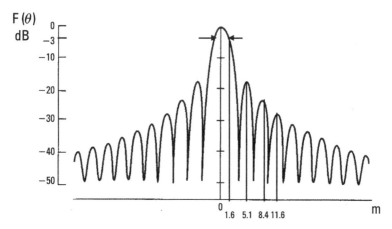

Figure 10.9 Radiation pattern.

We can then deduce from the table of the Bessel function $J_1(m)/m$ the radiation pattern of the circular aperture. This pattern is achieved in Figure 10.9, not in polar coordinates but in Cartesian coordinates, due to the very significant directivity of this radiating aperture, which would otherwise make the pattern unreadable.

According to the table $J_1(m)/m$:

- The 3-dB half-aperture angle is given by the formula $\sin\theta = 1.6\ \lambda/2\pi a$.
- The first secondary lobe is at −17.6 dB for $\sin\theta = 5.1\ \lambda/2\pi a$.
- The second secondary lobe is at −23.8 dB for $\sin\theta = 8.4\ \lambda/2\pi a$.
- The third secondary lobe is at −28 dB for $\sin\theta = 11.6\ \lambda/2\pi a$.

Table 10.3
Table of the Bessel function $J_1(m)/m$

x	$J_1(x)/x$	x	$J_1(x)/x$	x	$J_1(x)/x$
0.0	0.50000	5.1	−0.06610	10.2	−0.00065
0.1	0.49938	5.2	−0.06600	10.3	−0.00304
0.2	0.49750	5.3	−0.06528	10.4	−0.00533
0.3	0.49440	5.4	−0.06395	10.5	−0.00751
0.4	0.49007	5.5	−0.06208	10.6	−0.00955
0.5	0.48454	5.6	−0.05970	10.7	−0.01144
0.6	0.47783	5.7	−0.05687	10.8	−0.01316
0.7	0.46999	5.8	−0.05363	10.9	−0.01471

Table 10.3, (*Cont'd*)

0.8	0.46105	5.9	−0.05002	11.0	−0.01607
0.9	0.45105	6.0	−0.04611	11.1	−0.01724
1.0	0.44005	6.1	−0.04194	11.2	−0.01820
1.1	0.42809	6.2	−0.03757	11.3	−0.01896
1.2	0.41524	6.3	−0.03303	11.4	−0.01951
1.3	0.40156	6.4	−0.02838	11.5	−0.01986
1.4	0.38710	6.5	−0.02367	11.6	−0.02000
1.5	0.37196	6.6	−0.01894	11.7	−0.01994
1.6	0.35618	6.7	−0.01423	11.8	−0.01969
1.7	0.33986	6.8	−0.00959	11.9	−0.01924
1.8	0.32306	6.9	−0.00506	12.0	−0.01862
1.9	0.30587	7.0	−0.00067	12.1	−0.01783
2.0	0.28836	7.1	0.00354	12.2	−0.01688
2.1	0.27061	7.2	0.00755	12.3	−0.01579
2.2	0.25271	7.3	0.01131	12.4	−0.01457
2.3	0.23473	7.4	0.01481	12.5	−0.01324
2.4	0.21674	7.5	0.01803	12.6	−0.01180
2.5	0.19884	7.6	0.02095	12.7	−0.01029
2.6	0.18108	7.7	0.02355	12.8	−0.00871
2.7	0.16356	7.8	0.02582	12.9	−0.00707
2.8	0.14633	7.9	0.02774	13.0	−0.00541
2.9	0.12946	8.0	0.02933	13.1	−0.00373
3.0	0.11302	8.1	0.03057	13.2	−0.00205
3.1	0.09707	8.2	0.03146	13.3	−0.00039
3.2	0.08167	8.3	0.03202	13.4	0.00124
3.3	0.06687	8.4	0.03224	13.5	0.00282
3.4	0.05271	8.5	0.03213	13.6	0.00434
3.5	0.03925	8.6	0.03172	13.7	0.00578
3.6	0.02652	8.7	0.03100	13.8	0.00713
3.7	0.01455	8.8	0.03001	13.9	0.00838
3.8	0.00337	8.9	0.02875	14.0	0.00953
3.9	−0.00699	9.0	0.02726	14.1	0.01055
4.0	−0.01651	9.1	0.02554	14.2	0.01145
4.1	−0.02519	9.2	0.02363	14.3	0.01222
4.2	−0.03301	9.3	0.02155	14.4	0.01285
4.3	−0.03998	9.4	0.01932	14.5	0.01334
4.4	−0.04609	9.5	0.01697	14.6	0.01369
4.5	−0.05135	9.6	0.01453	14.7	0.01389
4.6	−0.05578	9.7	0.01202	14.8	0.01396
4.7	−0.05938	9.8	0.00947	14.9	0.01388
4.8	−0.06219	9.9	0.00691	15.0	0.01367

1.4.2 Isotropic Gain

We have seen in Chapter 1 that the isotropic gain of an antenna in a given direction was equal to

$$G_i = 4\pi U/P_e$$

where U is the radiation intensity in that direction and P_e is the power delivered to the antenna.

The modulus of the field at the point M in the main direction ($\theta = 0$) is, according to (10.10):

$$E_{oz} = \pi a^2 E_0/\lambda r$$

The power per unit area around the point M in this direction is

$$W = 1/2\varepsilon E^2 = 1/2\varepsilon\pi^2 a^4 E_0^2/\lambda^2 r^2$$

Therefore

$$U = Wr^2 = 1/2\varepsilon\pi^2 a^4 E_0^2/\lambda^2$$

The power provided to the aperture is

$$P_e = 1/2\varepsilon\pi^2 a^2 E_0^2$$

The isotropic gain of the aperture in the main direction is thus

$$G_i = 4\pi U/P_e = 4\pi^2 a^2/\lambda^2$$

$G_i = \pi^2 D^2/\lambda^2$, D being the paraboloid diameter.

1.5 Realization of a Parabolic Antenna

1.5.1 Illumination of the Entire Concave Surface of the Paraboloid by Spherical Waves

The paraboloid being of revolution around its axis, the pyramidal horn, with its rectangular aperture on its focus, must have a beam of revolution as well for covering the concave surface of the paraboloid, as indicated in the Figure 10.10, where the E-plane or H-plane patterns are identical. Therefore, the entire concave surface of the paraboloid is illuminated. In these conditions, knowing the diameter D of the paraboloid and the focal length f we can write the following formula:

$$\alpha = 2 \arctan \ (D/2f)$$

with α being the aperture angle of the beam. The aperture angle of each pattern must approximately be $\alpha/2$.

As we have seen Section 10.4, the parabolic antenna must be illuminated by spherical waves. These spherical waves, which by definition are electric fields of the same amplitude, in phase at each point, can only be effectively provided by a pyramidal horn within certain limits. Figure 10.11 shows a vertical section and an horizontal section of the horn with the spherical waves.

Let us take the example of the vertical section. Since the horn aperture is not concave, the electric fields on the aperture will be slightly out of phase with each other, with the maximum being at the edge of the horn; that is, a phase shift of

$$\varphi = 2\pi\Delta L/\lambda$$

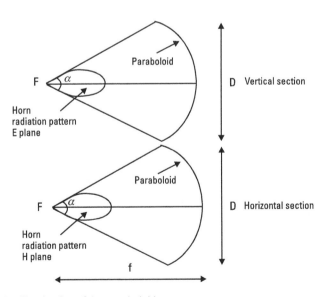

Figure 10.10   Illumination of the paraboloid.

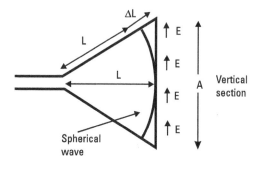

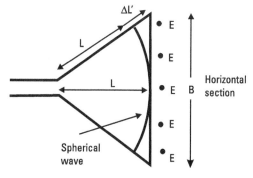

Figure 10.11   E- and H-plane sections of the horn.

with $\Delta L' = A^2/8L$ (geometry formula).

It is considered that for $\varphi <$ or equal to $\pi/4$, the field is uniform (same amplitude and same phase) on the aperture.

In the horizontal section we must, for the same reasons, have

$$\varphi = 2\pi\Delta L'/\lambda < \text{ or equal to } \pi/4$$

with $\Delta L' = B^2/8L$.

We must therefore verify that these two conditions are met, after calculating a pyramidal horn for a parabolic antenna.

1.5.2 Exercise: Parabolic Antenna Powered by Pyramidal Horn Calculation

The parabolic antenna would have a diameter of 20 cm, a focal length of 18 cm, and the frequency of use would be 22 GHz.

- Calculate the horn dimensions;
- Calculate the horn gain;
- Draw the radiation patterns for the E-plane and H-plane of the horn;
- Calculate the parabolic antenna gain;
- Calculate for the parabolic antenna, the aperture angle, the angle corresponding to the first secondary lobe and its level relative to the main lobe, and the same characteristics for the other two secondary lobes.

1.5.2.1 Horn Dimensions

According to the characteristics of the parabolic antenna the α angle can be calculated:

$$\alpha = 2 \text{ Arc tang. } (D/2f) = 58°$$

The aperture angle of the radiation pattern in the E-plane and H-plane must be in the order of $2\theta = \alpha/2 = 29°$. From the table in Figure 10.5 giving $F(\theta)$ with respect to x we find $x = 1.4$ for

$$F(\theta) = 0.7 \ (-3 \text{ dB})$$

with $x = \pi A \sin\theta/\lambda$. Hence $A = 1.4 \ \lambda/\pi \sin\theta = 0.025$ m.

It is then necessary to calculate the horn length L to have a maximum error of phase φ on the surface of the aperture less than $\pi/4$. We will take $\pi/8$.

We have $\varphi = 2\pi\Delta L/\lambda$ and $\Delta L = A^2/8L$. Hence $L = \pi A^2/4\lambda\varphi = 2A^2/\lambda = 0.089$ m.

As the paraboloid being of revolution around its axis, the aperture angle of the pattern in the H-plane must be the same as in the E-plane; that is $2\theta = 29°$.

From the table in Figure 10.6 giving $F(\theta)$ with respect to x we find $x = 1.8$ for

$$F(\theta) = 0.7 \ (-3 \text{ dB})$$

with $x = \pi B \sin\theta/\lambda$. Hence $B = 1.8\lambda/\pi \sin\theta = 0.032$m.

The horn length does not change; it is the phase error that will change. We then have:

$$\Delta L' = B^2/8 \ L = 0.0014 \ m$$

The phase error is

$$\varphi' = 2\pi\Delta L'/\lambda = 46°$$

which is $< \pi /4$. The result is thus the following:

$$A = 2.5 \text{ cm}$$
$$B = 3.2 \text{ cm}$$
$$L = 8.9 \text{ cm}$$

1.5.2.2 Horn Gain

The gain formula of the rectangular aperture must be applied:

$$G_i = 0.81 \ (4\pi \ AB/\lambda^2)$$

That is; $G_i = 11$ dB.

1.5.2.3 Horn Radiation Patterns

E-Plane

The characteristic function of the rectangular radiating aperture in the E-plane is (10.4):

$$F(\theta) = \sin \ (\pi \ A \ \sin\theta/\lambda)/(\pi \ A \ \sin\theta/\lambda)$$

If we change θ into $-\theta$, $F(\theta)$ does not change. Therefore, we just have to plot the curve for θ being between 0 and 90° and take its symmetric with respect to oz.

From the table in Figure 10.5 giving $F(\theta)$ as a function of x, we can draw this curve, knowing that $x = \pi \ A \ \sin\theta/\lambda$.

The pattern in the E-plane will be presented at the same time as the one in the H-plane.

H Plane

The characteristic function of the rectangular radiating aperture in the H-plane is (10.7):

$$F(\theta) = \frac{\pi^2}{4} \cos x \bigg/ \left(\frac{\pi^2}{4} - x^2 \right)$$

If we change θ into $-\theta$, $F(\theta)$ does not change. Therefore, we just have to plot the curve for θ being between 0° and 90° and take its symmetric with respect to oz. From Table 10.1 giving $F(\theta)$ as a function of x, this curve can be drawn, knowing that $x = \pi \ B \ \sin\theta/\lambda$.

We can note that in space, the radiation pattern of the pyramidal horn excited by a waveguide in TE_{10} mode is of revolution around the oz axis, and

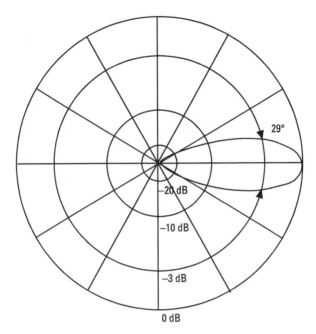

Figure 10.12   Horn pattern in the E-plane.

is therefore perfectly adapted for the total illumination of the paraboloid, itself revolving around *oz*. This is facilitated with the TE_{10} mode by the fact that the aperture angles of the radiation patterns in each plane are based only on one side of the radiating aperture rectangle:

E-plane: $\sin\theta = 1.4\lambda/\pi A$

H-plane: $\sin\theta = 1.8\lambda/\pi B$

To obtain the same angle, we need $B = 1.29\ A$.

1.5.2.4 Parabolic Antenna Gain

We have: $G_i = \pi^2 D^2/\lambda^2$; that is, $G_i = 33$ dB. To this gain we must add the horn gain 11 dBi. The total gain of the antenna is then:

$$G_i = 44 \text{ dB}$$

1.5.2.5 Characteristics of the Radiation Pattern of the Parabolic Antenna

The aperture angle level is −3 dB, given by $\sin\theta = 1.6\lambda/\pi D = 0.036$; hence $2\theta = 4°$.

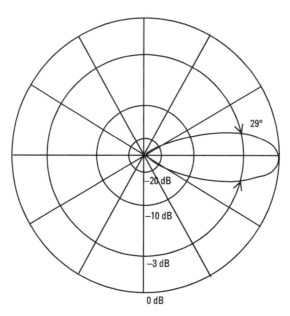

Figure 10.13   Horn pattern in the H plane.

The angle of the first secondary lobe level is −17.6 dB, given by $\sin\theta = 5.1\lambda - \pi D = 0.114$; hence $\theta = 6.5°$.

The angle of the second secondary lobe level is −23.8 dB, given by $\sin\theta = 8.4\lambda - \pi D = 0.187$; hence $\theta = 11°$.

The angle of the third secondary lobe level is −28 dB, given by $\sin\theta = 11.6\lambda - \pi D = 0.258$; hence $\theta = 15°$.

This aperture angle of 4° shows the care needed for pointing the two parabolic antennas of a microwave link. The technician would adjust the direction of each antenna for getting the maximum power level at the intermediate frequency (IF) of each receiver of the microwave link.

The radiation pattern given by the manufacturers is the envelope of the radiation (see Figure 10.15).

References

[1]   Stutzman, W. L., and G. A. Thiele, *Antenna Theory and Design*, Second Edition, Hoboken, NJ: Wiley, 1998.

[2]   Eyraud, L., G. Grange, and H. Ohanessian, *Théorie et Technique des Antennes* (in French), Vuibert, 1973.

Appendix 10A

Example of an Industrial Parabolic Antenna

Figure 10.14   A 37–40-GHz parabolic antenna.

Technical Specifications

- Frequency band: 37–40 G Hz.
- Vertical or horizontal polarization.
- Aperture angle: 2.8°.
- Gain:
  - Lower part of the band: 34.5 dBi.
  - Middle part of the band: 34.7 dBi.
  - Upper part of the band: 34.9 dBi.
- Impedance: 50 Ω.
- VSWR < 1.5.
- Front/back ratio: 60 dB.
- Paraboloid diameter: 0.2m.
- Maximum power: 500W.
- Height: 1612 mm.°

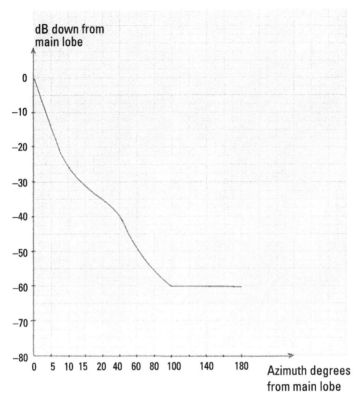

Figure 10.15   Radiation pattern envelope.

Part III:
Appendixes

Appendix A

Frequency Bands of Mobile Radio Networks

A.1 Frequency Regulations in the World

Radio frequencies are scarce resources. Many national interests make it hard to find common worldwide regulations.

The International Telecommunications Union (ITU) located in Geneva (Switzerland) is responsible for worldwide coordination of telecommunication activities (wired and wireless).

The ITU Radiocommunications sector (ITU-R) handles standardization in the wireless sector as well as frequency planning (formerly known as Consultative Committee for International Radiocommunication (CCIR)).

To have at least some success in worldwide coordination and to reflect national interests, the ITU-R has split the world into three regions:

1. Region 1 covers Europe, the Middle East, countries of the former Soviet Union, and Africa.

2. Region 2 includes Greenland and North and South America.

3. Region 3 comprises the Far East, Australia, and New Zealand.

Within these regions, national agencies are responsible for further regulations. These agencies include the Federal Communications Commission (FCC) in the United States and the European Conference for Posts and

Telecommunications (CEPT) in Europe in Monteux, Switzerland, in which 48 countries of region 1 are grouped.

To achieve at least some harmonization the ITU-R holds the World Radio Conference (WRC), to periodically discuss and decide frequency allocations for the three regions.

This is obviously a difficult task. It is the reason why some frequency bands are still different in each region.

Frequency bands handled by the CEPT are discussed below.

A.2 CEPT Frequency Allocations Used in Mobile Radio Networks for Mobile Communications

The frequency channels allocated by CEPT comprise two frequencies so that the networks can operate in duplex or half-duplex: frequency F_1 in the direction of base (or repeater) to mobile, and frequency F_2 in the direction of mobile to base (or repeater).

The spacing between these two frequencies, called duplex spacing, depends on the frequency band. The half-duplex is used with the push-to-talk operation in mobile PMR networks.

A.2.1 30–300-MHz Metric Band: VHF Band

The VHF band includes:

- 34–41-MHz band with channel spacing of 12.5 kHz and duplex spacing of 4.6 MHz;
- 68–88-MHz band with channel spacing of 12.5 kHz and duplex spacing of 4.6 MHz;
- 151–162-MHz band with channel spacing of 12.5 kHz and duplex spacing of 4.6 MHz.

These frequency bands are used only for PMR networks.

A.2.2 300–3000-MHz Decimeter Band: UHF Band

The UHF band includes:

- 380–400-MHz band with channel spacing of 12.5 kHz and duplex spacing of 10 MHz;

- 406–430-MHz band with channel spacing of 25 kHz and duplex spacing of 10 MHz;
- 440–470-MHz band with channel spacing of 25 kHz and duplex spacing of 10 MHz.

These frequency bands are only for PMR networks:

- 890–915-MHz band associated with the 935–960-MHz band, with channel spacing of 200 KHz and a duplex spacing of 45 MHz, for GSM 900;
- 1710–1785-MHz band associated with the 1805–1880-MHz band, with channel spacing of 200 KHz and a duplex spacing of 95 MHz, for GSM 1800;
- 1900–1980- , 2010–2025- , and 2110–2170-MHz bands, with channel spacing of 5 MHz, and duplex spacing of 190 MHz, for UMTS.

These frequency bands are only for operated public networks.

A.3 CEPT Frequency Allocations Used in Mobile Phone Networks for Connections by Microwave Links between the Antenna Site and the Switching Center or for Links between Sites

The frequencies allocated by CEPT in this case are in the spectrum bands given below.

A.3.1 3–30-GHz Centimetric Band: SHF Band

The superhigh frequency (SHF) band includes:

- 12.75–13.25-GHz band with channel spacing of 3.5 MHz for a bit rate of 2×2 MBit/sec;
- 22–23.6-GHz band with channel spacing of 3.5 MHz for a bit rate of 2×2 MBit/sec;
- 24.5–26.6-GHz band with channel spacing of 3.5 MHz for a bit rate of 2×2 MBit/sec.

A.3.2 30–300-GHz Millimetric Band: EHF Band

The extrahigh frequency (EHF) band includes:

- 37–39.5-GHz band with channel spacing of 3.5 MHz for a bit rate of 2×2 MBit/sec.

References

[1]   Schiller, J. (ed.), *Mobile Communications*, Second Edition, Boston: Addison Wesley, 2003.

[2]   National Agency of Radioelectric Frequencies (ANFR), Tableau National de Repartition des Bandes de Fréquences (National Table of Frequency Spectrum), 2008.

Appendix B

Vector Calculus Reminders

B.1 Vectors and Scalars

B.1.1 Vectors

The vector is a mathematical object with a module, a direction, and an orientation in space (Figure B.1). Examples are strength, speed, acceleration, electric field, and magnetic field.

B.1.2 Scalar

The scalar is a quantity or magnitude without direction. Examples are mass, length, time, temperature, and potential.

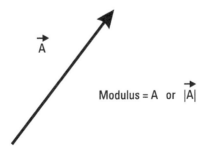

$\vec{A}$

Modulus = A or $|\vec{A}|$

Figure B.1   Vector.

B.1.3 Vector Algebra

Two vectors $\vec{A}$ and $\vec{B}$ are equal if they have the same module, the same direction, and the same orientation. A vector of the opposite orientation to $\vec{A}$ but having the same module and the same direction is noted as $-\vec{A}$.

We can make the addition or subtraction of two vectors $\vec{A}$ and $\vec{B}$. This process is called a geometric sum (Figure B.2).

The product of a vector $\vec{A}$ by a scalar m is a vector $\mathbf{m\vec{A}}$.

B.1.4 Unit Vectors: Direct Orthogonal System

A unit vector is a vector of modulus 1. An application would be a coordinate system in three dimensions: ox, oy, oz (Figure B.3). This system is a direct orthogonal coordinate if a corkscrew that turns from $0x$ to $0y$ sinks in the positive direction of oz. For example, $\vec{\mathbf{i}}$, $\vec{\mathbf{j}}$, $\vec{\mathbf{k}}$, are unit vectors.

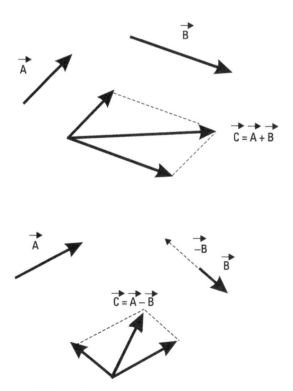

Figure B.2   Vector addition and subtraction.

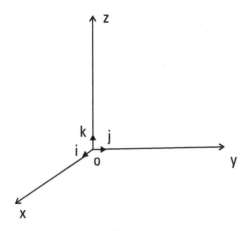

Figure B.3 A coordinate system in three dimensions.

B.1.5 Representation of a Vector in The System of Axes

Let $\vec{V}$ be a vector represented in a system of axes ox, oy, oz, by its projections on the three axes V_1, V_2, V_3.

We can write:

$$\vec{V} = V_1\vec{i} + V_2\vec{j} + V_3\vec{k}$$

where V_1, V_2, V_3 are the modules of vectors $\vec{V}_1$, $\vec{V}_2$, $\vec{V}_3$.

If the vector $\vec{V}$ is a function of the coordinates x, y, z from its origin, we say that we have a vector function of point $\vec{V}$ (x,y,z); for example, velocity at each point of a fluid.

B.2 Scalar Product and Vector Product

B.2.1 Scalar Product

B.2.1.1 Definition

The scalar product of two vectors $\vec{A}$ and $\vec{B}$ (read A scalar B) is defined as the product of the modules A and B multiplied by the cosine of the angle they form (Figure B.4):

$$C = AB\cos\theta$$

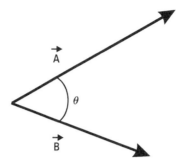

Figure B.4   Scalar product.

This is a scalar and not a vector.

B.2.1.2 Scalar Product Property

If $\vec{A} = A_1\vec{i} + A_2\vec{j} + A_3\vec{k}$ and $\vec{B} = B_1\vec{i} + B_2\vec{j} + B_3\vec{k}$, we have:

$$\vec{A}.\vec{B} = A_1B_1 + A_2B_2 + A_3B_3$$

B.2.2 Vector Product

B.2.2.1 Definition

The vector product of $\vec{A}$ and $\vec{B}$ is the vector $\vec{C} = \vec{A} \wedge \vec{B}$. The modulus **C** is the product of the modules A and B multiplied by the sine of the angle they form:

$$C = AB \sin\theta$$

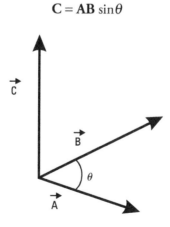

Figure B.5   Vector product.

The direction of the vector $\vec{C}$ is perpendicular to the plane of $\vec{A}$ and $\vec{B}$, and its orientation is given by the rule of the corkscrew (Figure B.5).

B.2.2.2 Vector Product Property

If $\vec{A} = A_1\vec{i} + A_2\vec{j} + A_3\vec{k}$ and $\vec{B} = B_1\vec{i} + B_2\vec{j} + B_3\vec{k}$, we can write $\vec{A} \wedge \vec{B}$ in the form of a determinant:

$$\vec{A} \wedge \vec{B} = \begin{vmatrix} \vec{i} & \vec{j} & \vec{k} \\ A_1 & A_2 & A_3 \\ B_1 & B_2 & B_3 \end{vmatrix}$$

By developing, we obtain:

$$\begin{vmatrix} \vec{i} & \vec{j} & \vec{k} \\ A_1 & A_2 & A_3 \\ B_1 & B_2 & B_3 \end{vmatrix} = \vec{i}\begin{vmatrix} A_2 & A_3 \\ B_2 & B_3 \end{vmatrix} + \vec{j}\begin{vmatrix} A_1 & A_3 \\ B_1 & B_3 \end{vmatrix} + \vec{k}\begin{vmatrix} A_1 & A_2 \\ B_1 & B_2 \end{vmatrix}$$

which gives:

$$\vec{A} \wedge \vec{B} = \vec{i}(A_2B_3 - A_3B_2) + \vec{j}(A_1B_3 - A_3B_1) + \vec{k}(A_1B_2 - A_2B_1)$$

B.3 Curl

We must first define the differential and vector operator NABLA: $\vec{\nabla}$

$$\vec{\nabla} = d/dx\,\vec{i} + d/dy\,\vec{j} + d/dz\,\vec{k}$$

The curl of a vector $\mathbf{V}$ is the vector product: $\vec{\nabla} \wedge \vec{V}$.

Let $\vec{V}$ be a vector function of a point in a system of rectangular axes ox, oy, oz:

$$\vec{V}(x, y, z,) = V_1\vec{i} + V_2\vec{j} + V_3\vec{k} \quad \text{(see Section B.1.5)}$$

Therefore the curl of $\vec{V}$ (x,y,z) is

$$\vec{\nabla} \wedge \vec{V} = \begin{vmatrix} \vec{i} & \vec{j} & \vec{k} \\ d/dx & d/dy & d/dz \\ V_1 & V_2 & V_3 \end{vmatrix}$$

$$= (dV_3/dy - dV_2/dz)\vec{i} + (dV_1/dz - dV_3/dx)\vec{j} + (dV_2/dx - dV_1/dy)\vec{k}$$

The curl of the vector $\vec{V}$ (x,y,z) is perpendicular to the vector $\vec{V}$. Using fluid as an example, the curl of the speed of a river flow can rotate a turbine.

B.4 Stokes' Theorem

Along a closed curve C, the circulation of the vector $\vec{A}$ tangential to C is equal to the flow of the curl of $\vec{A}$ through any surface S based on the outline C.

$$\int_C \vec{A}.\,\vec{dl} = \iint_S \vec{\nabla} \wedge \vec{A}.\,dS$$

References

[1]   Leonov, S. A., and A. I. Leonov, *Mathematical Handbook for Electrical Engineers*, Norwood, MA: Artech House, 2004.

[2]   Spiegel, M. R., *Vector Analysis*, New York: McGraw-Hill, 1973.

Appendix C

Complex Number Reminders

C.1 Definitions

A complex number $\overline{A}$ is a combination of two real numbers a and b written in the form:

$$\overline{A} = a + jb \tag{C.1}$$

in which

$$j = \sqrt{-1}$$

Therefore, j is an imaginary number. The variable a is called the real part and the variable b is called the imaginary part.

The modulus of the complex number $\overline{A}$ is a positive number:

$$A = \sqrt{a^2 + b^2}$$

The argument of the complex number $\overline{A}$ is the angle $\alpha = \text{Arctan } b/a$.

We can thus represent the complex number in a plane xoy by a point P of coordinates a and b, the segment OP forming an angle α with ox (Figure C.1).

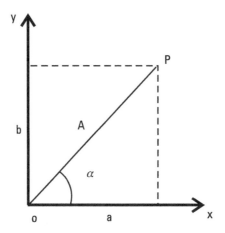

Figure C.1 Representation of a complex number in a complex plane.

The plane xoy is called the complex plane, the axis ox is the real axis, and the axis oy is the imaginary axis.

This representation shows that (C.1) can also be written as

$$\overline{A} = A(\cos\alpha + j\sin\alpha) \tag{C.2}$$

or

$$\overline{A} = Ae^{j\alpha} \tag{C.3}$$

according the Euler formula $e^{j\alpha} = \cos\alpha + \sin\alpha$ (see Section C.3).

The relations (C.1), (C.2), and (C.3) are the three imaginary notations allowing to represent a complex number in a complex plane.

C.2 Equality of Two Complex Numbers

Let there be two complex numbers:

$$\overline{A}_1 = a_1 + jb_1$$

$$\overline{A}_2 = a_2 + jb_2$$

If

$$A_1 = A_2$$

we have

$$a_1 = a_2$$
$$b_1 = b_2$$

In other words, there is equalization of real parts on one hand, and of imaginary parts on the other hand.

C.3 Euler's Formula

Euler's formula allows to connect $e^{j\alpha}$ to $\cos\alpha$ and to $\sin\alpha$:

$$e^{j\alpha} = \cos\alpha + j\,\sin\alpha \tag{C.4}$$

We can deduce the following formulas from this formula:

$$e^{-j\alpha} = \cos\alpha - j\,\sin\alpha$$
$$\cos\alpha = (e^{j\alpha} + e^{-j\alpha})/2$$
$$\sin\alpha = (e^{j\alpha} - e^{-j\alpha})/2j$$

C.4 Representation of a Sine Function by an Exponential Function

By replacing α by ωt in (C.4), we find:

$$Ae^{j\omega t} = A(\cos\omega t + j\,\sin\omega t)$$

According to Section C.2, we can equalize the real parts on one hand, and the imaginary parts on the other hand. We can thus replace: $y = A\,\sin\omega t$ by $\bar{y} = Ae^{j\omega t}$.

Similarly, if we have the function $y = A\,\sin(\omega t - \varphi)$, we can write:

$$\overline{y} = Ae^{j(\omega t - \varphi)} = Ae^{j\omega t}e^{-j\varphi}$$

We will often use this formula because it is perfect to separate the phase term from the time term.

Reference

[1]   Leonov, S. A., and A. I. Leonov, *Mathematical Handbook for Electrical Engineers*, Norwood, MA: Artech House, 2004.

Appendix D

Electrostatic Reminders

D.1 Coulomb's Law

Two electric charges, m and m', exert on each other a force directed along a straight line that joins them and is equal to

$$\mathbf{F} = m \times m' / 4\pi\varepsilon r^2$$

The force $\mathbf{F}$ repels or attracts charges according to their signs (charges of the same sign repel each other). ε is the dielectric constant of the medium and r the distance between the charges. The reference of the dielectric is the vacuum whose dielectric constant is ε_0 : $\varepsilon_0 = 8.842 \ 10^{-12}$ Farad/m in **MKSA** units, and there is a relation between ε and ε_0 : $\varepsilon = \varepsilon_r \varepsilon_0$.

If the air is the dielectric, this is what we are interested in, because the medium around the antenna is the latter, $\varepsilon_r = 1 + 5.67 \ 10^{-4}$.

In this case the difference between ε and ε_0 is negligible.

D.2 Electric Field

The electric field $\vec{E}$ created by an electric charge m at a point P in space at a distance r is defined as follows:

- Its origin is the point.
- Its direction is that of the force that would be exerted on any electric charge placed at that point.
- Its orientation and module are those of the force that would be exerted on a positive charge $+ 1$ (Figure D.1).

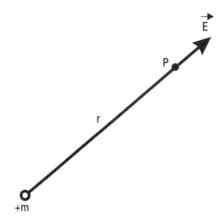

Figure D.1   Electric field.

Its modulus is thus

$$E = m/4\pi\varepsilon r^2$$

We will see later that E is expressed in volt/meter.

D.3 Conductors and Insulators

Some materials rubbed with a piece of cloth carry charges only in the place that has been rubbed. These are called insulators or dielectrics. Examples include glass, Teflon, ebonite, resin, mica, and air.

Other materials held in the hand by an insulating handle and rubbed under the same conditions will carry charges evenly distributed over the entire surface in a free state. These are called conductors. Examples include copper, brass, silver, and gold.

D.3.1 Conductors

D.3.1.1 General Characteristics

The charges evenly distributed within an insulated conductor in the previous experience are electrons disconnected from their atoms. These electrons are negative charges, leaving in the atom positive charges attached to the atom nucleus (protons). These charges in the free state are in equilibrium in the conductive surface if that surface is insulated. There is no electric field inside the conductor.

If we establish a potential difference between terminals A and B of a conductive wire, free charges will also appear uniformly and move from the lower potential to the higher potential. There is an electric current and an electric field inside the conductor. This current is called a conduction current. The relation between current and charges is $i = dq - dt$.

D.3.1.2 Perfect Conductor/Good Conductor

The perfect conductor is characterized by an infinite conductivity σ, which is expressed in mho/m. It is characterized by the absence of any electromagnetic phenomenon. There is, therefore, no electric field tangential to a perfect insulated conductor.

The perfect conductor does not exist, but the measurements show that the "good conductors," such as aluminum ($\sigma = 3.39 \ 10^7$), copper ($\sigma = 5.65 \ 10^7$), and silver ($\sigma = 6.1 \ 10^7$) come close to being perfect conductors when we are interested in phenomena at high frequency (>100 kHz), because there is also an electromagnetic vacuum and thus a tangential field at a high frequency equal to zero.

D.3.2 Dielectric: First Maxwell Equation

Let us take the example of the plane capacitor whose dielectric is the air and that is placed in a closed circuit comprising a battery of electromotive force e, and a switch (Figure D.2). When closing the switch, a current i flows through the conductor wire. The current i consists of free charges, uniformly distributed in the conductor, which are going to be deposited on the plates of the plane capacitor. These charges create an electric field, $\vec{E}$, that is evenly distributed at each point of the dielectric and is tangent to lines called field lines (Figure D.3).

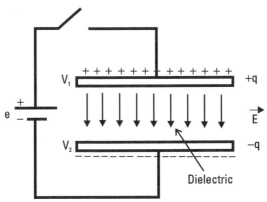

Figure D.2 Plane capacitor.

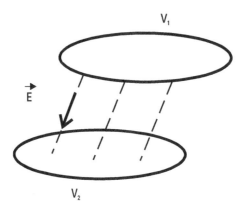

Figure 4.3   Field lines.

On a field line of length l we have the relation:

$$V_1 - V_2 = \int_0^l \vec{E}.\,\vec{dl}$$

In the MKSA system, the electric field is thus expressed in volt/m. The potential difference V_1-V_2 creates in the dielectric a displacement current generated by the displacement of the electrons present in the atoms of the dielectric. This current, which follows the field lines, is defined, at each point of the dielectric as the current $\vec{j}$ flowing across an elementary surface surrounding this point. It is expressed in amperes per square meter in the MKSA system. If the potential difference is sinusoidal (the two plates being fed by a sinusoidal source), the displacement current is sinusoidal.

There is a relationship between the current and the electric field:

$$\vec{j} = \varepsilon d\vec{E}/dt$$

This is the first Maxwell equation, applicable only to the dielectric.

D.4 Electrostatic Energy

Per unit area of dielectric traveled, the electrostatic energy is

$$W = 1/2 \; \varepsilon E^2$$

W is expressed in Watt/m^2 and E in volt/m.

References

[1]   Rocard, Y., *Electricité* (in French), Paris: Masson et Cie., 1956.

[2]   Grivet, P., *Electrostatique—Magnetisme* (in French), Paris: Paris University, 1955.

Appendix E

Electromagnetism Reminders

E.1 Coulomb's Law

Two magnetic masses, m and m', located at a distance r from each other repel or attract each other with a force:

$$F = m \times m' / 4\pi\mu r^2$$

When the magnetic masses are of the same sign they repel each other. μ is the magnetic permeability of the medium and r is the distance between the masses. The reference of the medium is the vacuum whose the magnetic permeability is μ_0:

$$\mu_0 = 1.256637 \ 10^{-6} \text{ Henry/m in the MSA system.}$$

There is a relation between μ and μ_0 : $\mu = \mu_r\mu_0$. If the air is the medium, what we are interested to, because the latter is around the antenna, $\mu_r = 1 + 3.7 \ 10^{-7}$.

In this case the difference between μ and μ_o is negligible.

E.2 Magnetic Field

The magnetic field $\vec{H}$ created by a magnetic charge m at a point P in space at a distance r is defined as follows:

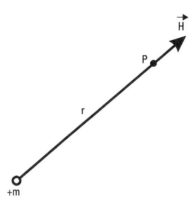

Figure E.1   Magnetic Field

- Its origin is the point.
- Its direction is the direction of the force that would be exerted on any magnetic charge placed at that point.
- Its orientation and module are those of the force that would be exerted on a positive charge + 1 (Figure E.1).

Its module is thus:

$$H = m/4\pi\mu r^2$$

The magnetic field is expressed in ampere/meter (see next section).

E.3 Laplace Law

The magnetic field created by a conductor element of length dl, travelled by a current i, at a point M (Figure E.2) is:

$$d\vec{H} = \vec{i}dl \sin\alpha/4\pi r^2 \tag{E.1}$$

This field is perpendicular to the plane defined by the conductor and the point M and is directed to the left of the ampere observer looking at the point M.

In the MKSA system, the magnetic field is thus expressed in amperes/m.

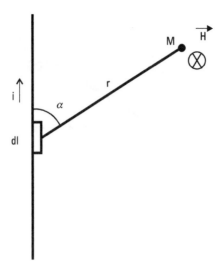

Figure E.2 Laplace Law

E.4 Vector Potential

If l is the length of the conductive wire, the vector $\vec{A}$ is called vector potential at a point M if:

$$\vec{A} = \int_l \vec{i}\, dl/4\pi r$$

That is:

$$d\vec{A} = \vec{i}dl/4\pi r$$

Let us represent the previous diagram (Figure E.2) in the plane $y0z$ of a coordinate system in 3 dimensions $0x$, $0y$, $0z$ (Figure E.3).

Let us calculate the Curl of $d\vec{A}$.

The $d\vec{A}$ coordinates are:

$$d\vec{A} \begin{vmatrix} dA_x = 0 \\ dA_y = 0 \\ dA_z = idl/4\pi r \end{vmatrix}$$

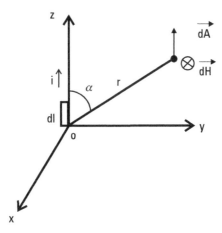

Figure E.3   Potential Vector

We have thus:

$$\vec{\nabla} \wedge d\vec{A} = \begin{vmatrix} i & j & k \\[4pt] \dfrac{d}{dx} & \dfrac{d}{dy} & \dfrac{d}{dz} \\[8pt] 0 & 0 & idl/4\pi r \end{vmatrix}$$

We find:

$$\vec{\nabla} \wedge d\vec{A} = \vec{i}\,d(idl/4\pi r)/dy$$

The module of the Curl is:

$$\left|\vec{\nabla} \wedge d\vec{A}\right| = (idl/4\pi r^2)dr/dy = (idl/4\pi r^2)y/r \text{ because } r = \sqrt{x^2 + y^2}.$$

As $y/r = \sin\alpha$, we can write [see (E.1)]:

$$\left|\vec{\nabla} \wedge d\vec{A}\right| = idl\sin\alpha/4\pi r^2$$

We have thus [see (E.1)]:

$$\left|\vec{\nabla} \wedge d\vec{A}\right| = d\vec{H}$$

Therefore:

$$|\vec{\nabla} \wedge \vec{A}| = \vec{H}$$

(E.2)

At each point the vector potential created by a HF current in a conducting wire is parallel to the conductive wire and its Curl is equal to the magnetic field created at that point.

The field $\vec{H}$ is said to derive from a vector potential $\vec{A}$.

E.5 Maxwell-Faraday Formula—2nd Maxwell Equation

Let a copper wire being a closed immobile circuit C of surface S in variable fields $\vec{H}$ perpendicular to the circuit (Figure E.5).

According to Lenz's Law:

$$e = -d\Phi/dt = -d/dt \iint_S \vec{B} \cdot \vec{dS}$$

(E.3)

where e is the electromotive force developed in the circuit, Φ the flux across the surface S of the circuit and $\vec{B}$ the induction vector, related to the magnetic field by the following relation: $\vec{B} = \mu\vec{H}$ (μ permeability of the air).

We can also write:

$$e = \int_C \vec{E} \cdot dl = \iint_S \vec{\nabla} \wedge \vec{E} \cdot \vec{dS}$$

(E.4)

according to the Stokes's formula (see Appendix B).

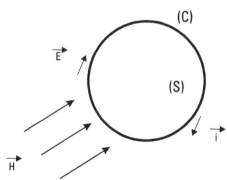

Figure E.4 Lenz's Law

We then have, by comparing the equations E.3 and E.4:

$$\vec{\nabla} \wedge \vec{E} = -d\vec{B}/dt \qquad (E.5)$$

This is the 2nd Maxwell equation.

By substituting in (E.5) $\vec{B}$ by $\mu\vec{H}$, this equation shows that if there is an electric field at a point in space, there is also a perpendicular magnetic field and vice versa.

E.6 Maxwell–Ampere Relation—3rd Maxwell Equation

Let us suppose that we have, in a dielectric, a conductive vertical wire travelled by an alternating current i (Figure E.5)

According to Laplace's law, an element dl creates at a point M of the dielectric a magnetic field perpendicular to the plane formed by the wire and the point M.

From the conclusions of the previous paragraph, there is necessarily at M an electric field $\vec{E}$ perpendicular to the magnetic field. This field is in the plane of the point M and of the wire. The same thing happens for all the points M in the same plane. We thus obtain electric field lines in this plane (see Appendix D). Displacement currents $\vec{j}$ follow these field lines.

Maxwell demonstrated that there was between the magnetic field and the displacement current the relation:

$$\vec{\nabla} \wedge \vec{H} = \vec{j}$$

This is the 3rd Maxwell equation.

E.7 Proportionality Relation—4th Maxwell Equation

The formula $\vec{\mathbf{B}} = \mu\vec{H}$ cited above, is the 4th Maxwell equation.

E.8 Electromagnetic Energy

The same way as in electrostatic, the electromagnetic energy per travelled unit area is:

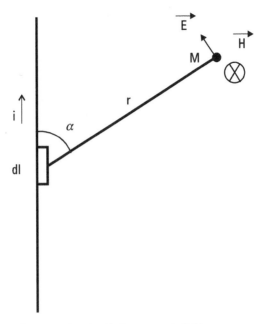

Figure E.5   Electric field associated with the magnetic field

$$W = \mu H^2 / 2$$

W is expressed in Watt/m^2 and H in amp/m in the MKSA system.

References

[1]   Rocard, Y., *Electricité* (in French), Paris: Masson et Cie., 1956.

[2]   Grivet, P., *Electrostatique—Magnetisme* (in French), Paris: Paris University, 1955.

Appendix F

Physics of Vibration Reminders

F.1 Wave Definition

A wave is the propagation of a vibratory disturbance in an elastic medium such as air (sound pipe), water (stone thrown into a pond, sea waves caused by the wind that varies the height of the water), string (string stretched between two poles, piano, violin, guitar), spring, electromagnetic waves, and so forth. Indeed, in an elastic medium any volume element undergoing deformation is capable of transmitting energy to a neighboring element. The wave does not carry material, but only energy. Most often the wave is sinusoidal, or if it is not, it is convertible into sinusoidal functions (Fourier series).

There are two types of waves:

1. Transverse waves, where the points of the propagation medium move perpendicularly to the direction of the disturbance. Examples include sea waves, a stone in a pond, the waves of earthquakes, and electromagnetic waves due to the electric charges' oscillation in the vicinity of an antenna.

2. Longitudinal waves, where the points of the propagation medium move in the direction of propagation. Examples include the compression or decompression of a spring and the sound in a sound pipe.

F.2 Wave Equation

Let us assume that the wave moves in the direction oz. This wave can be modeled by the differential equation of second order:

$$d^2u/dz^2 - 1/c^2\, d^2u/dt^2 = 0 \qquad\qquad (F.1)$$

where u is the amplitude of the wave and c its propagation speed. This equation has the following general solution:

$$U(z,t) = f_1(t - z/c) + f_2(t + z/c)$$

where f_1 is a wave propagating in the direction of the positive z, since the state of the point $z = a$ at time $t = a/c$ is the same $f_1(0)$ than the state of the point $z = 0$ at time $t = 0$.

The term f_2 is a wave that propagates in the opposite direction (towards the negative z). The wave f_1 is called direct wave and the wave f_2 is the reflected wave. In the event that $f_2 = 0$, there is no reflected wave. The wave is said to be traveling because it is not stopped by a reflection.

If f_2 is not null, the wave is reflected on an obstacle and we obtain a standing wave.

We will successively study the properties of the traveling wave and the standing wave.

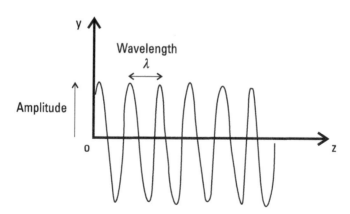

Figure F.1   Traveling wave.

F.3 The Traveling Wave Equation

A traveling wave is a wave moving forward in space. A transverse sinusoidal traveling wave is represented in Figure F.1.

The vibration source is placed at o and the sinusoidal vibrations are made according to oy:

$$y = a \sin \omega t \tag{F.2}$$

The wave is characterized by its amplitude, a, its speed, c, and its wavelength, λ, where the wavelength is the distance traveled by the wave during a period T of the vibration signal.

We can then write the formula :

$$\lambda = cT \tag{F.3}$$

The solution of the wave equation for the traveling wavecan be written as:

$$u(z,t) = f_1(t - z/c)$$

Let us calculate the wave amplitude $u(z,t)$ in a point M on the oz axis at the distance z of o. Due to the propagation of the vibration along oz, the amplitude in M at time t is created by the vibration at time $t - \dfrac{r}{c}$.

The (F.2) can be written as (see Appendix C):

$$\bar{y} = ae^{j\omega t} \tag{F.4}$$

Therefore the wave amplitude in M at the instant t is

$$\bar{u}(z,t) = ae^{j\omega(t-z/c)}$$

which is of the form $f_1(t - z/c)$.

We have the relation $T = 2\pi/\omega$, thus $c = \lambda\omega/2\pi$ [see (F.3)]. Therefore:

$$\bar{u}(z,t) = ae^{(j\omega t - 2\pi z/\lambda)}$$

which can be written as (see Appendix C):

$$\bar{u}(z,t) = a\sin(\omega t - 2\pi z/\lambda) \tag{F.5}$$

This is the equation of the traveling wave. This relation means that when the wave has traveled distance z, it has undergone a phase shift (delay) of $2\pi z/\lambda$ radians.

To simplify, $2\pi/\lambda$ is often replaced by β, which is very rightly called the phase constant ($\beta = 2\pi/\lambda$). β is expressed in radian/meter.

F.4   The Standing Wave Equation

A standing wave is a wave that oscillates without moving. It is the result of the combination of a traveling wave with its own reflection on an obstacle. We have represented in Figure F.2 the wave pathway, from the vibration source S, reflecting on an obstacle.

According to the theory of images, it is as if there was a second source S' in phase opposition, symmetric with respect to the obstacle.

With $\beta = 2\pi/\lambda$, the vibration supplied by the source S at a point M at a distance z from the obstacle will have as an equation:

$$\bar{u}_1(z,t) = ae^{j\omega t}e^{-j\beta(d-z)}$$

The vibration provided by the source S' at a point M will have as an equation:

$$\bar{u}_2(z,t) = -ae^{j\omega t}e^{-j\beta(d+z)}$$

The resulting vibration $u(z,t)$ at a point M will thus have as an equation:

$$\bar{u}(z,t) = u_1 + u_2 = ae^{j\omega t}e^{-j\beta d}\left(e^{j\beta z} - e^{-j\beta z}\right)$$

which can be written, according to the formula $\sin x = e^{jx} - e^{-jx}/2j$, and by replacing β by $2\pi/\lambda$.

$$\bar{u}(z,t) = 2ja\sin 2\pi z/\lambda\; e^{j(\omega t - 2\pi d/\lambda)} \qquad (F.6)$$

This is the equation of the standing wave.
We can therefore deduce:

- That each point is animated by a sinusoidal movement of the same frequency as the source and that this movement has the same phase

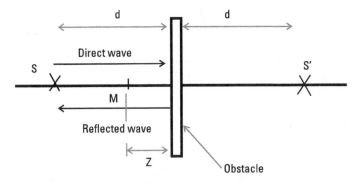

Figure F.2 Direct wave: reflected wave.

at all points, since the phase shift $2\pi\, d/\lambda$ does not depend on z. The wave does not move, which is why it is called a standing wave. That the amplitude depends on z, since it is equal to $2a\, \sin 2\pi z/\lambda$.

It is zero for all the points of abscissa z_n, such as

$$2\pi z_n/\lambda = k\pi$$

That is to say:

$$z_n = k\lambda/2$$

That is, for all the points N located at a distance from the obstacle equal to an integer of $\lambda/2$ or to an even number of $\lambda/4$. These are the nodes.

The amplitude is maximum and equal to 2a for all the points of abscissa z_A, such as

$$2\pi z_A/\lambda = (2k + 1)\pi/2$$

That is to say:

$$z_A = (2k + 1)\lambda/4$$

That is, for all the points A located at a distance from the obstacle equal to an odd number of $\lambda/4$. These are the antinodes.

Figure F.3 represents the standing wave. We can see that the nodes do not move on oz and that the antinodes take successively the positions 1, 2, 3, 4, 5, 4, 3, 2, 1 during a period of $e^{j\omega t}$. This is how the current varies on a half-wave antenna, reversing every half period.

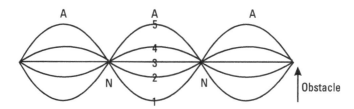

Figure F.3   Standing wave.

F.5   Radiation of Point Sources Aligned with the Same Spacing and in Phase

Let us consider a series of point sources, radiating in phase balance in all directions and placed vertically on a segment of length D at the same distance to each other. We will study the influence of these sources at a distant point P and in the plane shown in Figure F.4. This scheme is valid for radioelectricity, acoustics, and optics.

Let x be the distance between the midpoint of the segment and any source. At a remote point P from the segment, the path difference between the radiation of this source and the radiation from the central source is

$$OH = x \sin\theta$$

Let the radiation from each source be

$$\bar{a} = Ae^{j\omega t}$$

The instantaneous value of the radiation at the point P of the central source located in O can be written as

$$\bar{y}_0 = Ae^{j\omega t}e^{-j\beta r}$$

where r is the distance OP, β is the phase constant of the wave $\beta = 2\pi/\lambda$, and A is the amplitude of the radiation from the source.

The radiation at the point P of n sources located in a small segment dx around M at the distance x of O will be the in phase lead of $\beta x \sin\theta$. The instantaneous value of this radiation will thus be

$$\bar{y}_M = nAe^{j\omega t}e^{-j\beta r}e^{j\beta x \sin\theta}$$

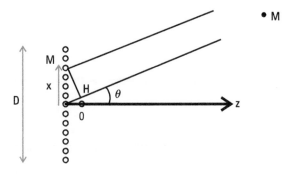

Figure F.4 Radiation in the vertical plane of a series of point sources vertically aligned with the same spacing and in phase.

The instantaneous radiation in P of the N sources of the segment will thus be

$$\bar{y} = nAe^{j\omega t}e^{-j\beta r}\int_{-D/2}^{+D/2}e^{j\beta x\sin\theta}dx$$

The result of the integral is

$$(e^{j\beta\frac{D}{2}\sin\theta} - e^{-j\beta\frac{D}{2}\sin\theta})/j\beta\sin\theta = 2\sin(\beta D/2\,\sin\theta)/\beta\sin\theta$$

because: $\sin\alpha = (e^{j\alpha} - e^{-j\alpha})/2\,j$ (see Appendix C).

The instantaneous total radiation at the point P is, in replacing β by $2\pi/\lambda$:

$$y = nA\big[\sin(\pi D\sin\theta/\lambda)/\,(\pi\sin\theta/\lambda)e^{j\omega t}e^{-j\beta r}$$

By multiplying the numerator and denominator by D and replacing nD with N, total number of sources in the segment, we obtain:

$$\bar{y} = NA\big[\sin(\pi D\sin\theta/\lambda)/(\pi D\sin\theta/\lambda)e^{j\omega t}e^{-j\beta r}$$

The characteristic function allowing to plot the radiation pattern in polar coordinates in the vertical plane is

$$F(\theta) = \sin\,(\pi D\,\sin\theta/\lambda)/(\pi D\,\sin\theta/\lambda) \tag{F.3}$$

This function is thus of the form $y = \sin x/x$. Its curve is represented in Figure F.5.

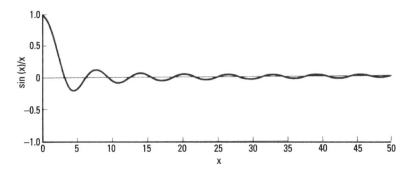

Figure F.5   Curve sin x/x.

Tables F.1 and F.2 allow us to plot the radiation pattern in the vertical plane (Figure F.6). The diagram in the vertical plane is necessarily symmetrical around the segment and the axis oz, so we just have to vary θ between 0 and $\pi/2$.

To draw the radiation pattern, we just have to search Table F.1 for the values of θ, ranging from 0 to $\pi/2$, for which $F(\theta) = \sin x/x$ is maximum or zero. We thus obtain Table F.2.

The pattern in the vertical plane shows that we obtain the following:

- Two important lobes pointing in the direction of the positive x and negative x and inscribed in an angle $\alpha = \text{Arc} \sin \lambda/D$.

- Eight secondary lobes.

- A gain in the direction $\theta = 0$ equal to the number N of sources (see (F.3) for $f(\theta) = 1$)).

Table F.1
sin x/x

x	$\sin(x)/x$	x	$\sin(x)/x$	x	$\sin(x)/x$
0.0	1.00000	5.1	−0.18153	10.2	−0.06861
0.1	0.99833	5.2	−0.16990	10.3	−0.07453
0.2	0.99335	5.3	−0.15703	10.4	−0.07960
0.3	0.98507	5.4	−0.14310	10.5	−0.08378
0.4	0.97355	5.5	−0.12828	10.6	−0.08705
0.5	0.95885	5.6	−0.11273	10.7	−0.08941
0.6	0.94107	5.7	−0.09661	10.8	−0.09083
0.7	0.92031	5.8	−0.08010	10.9	−0.09132
0.8	0.89670	5.9	−0.06337	11.0	−0.09091
0.9	0.87036	6.0	−0.04657	11.1	−0.08960
1.0	0.84747	6.1	−0.02986	11.2	−0.08743

Table F.1 Cont'd.

x	sin(x)/x	x	sin(x)/x	x	sin(x)/x
1.1	0.81019	6.2	−0.01340	11.3	−0.08443
1.2	0.77670	6.3	−0.00267	11.4	−0.08064
1.3	0.74720	6.4	−0.01821	11.5	−0.07613
1.4	0.70389	6.5	0.03309	11.6	−0.07093
1.5	0.66500	6.6	0.04720	11.7	−0.06513
1.6	0.62473	6.7	0.06042	11.8	−0.05877
1.7	0.58333	6.8	0.07266	11.9	−0.05194
1.8	0.54103	6.9	0.08383	12.0	−0.04471
1.9	0.49805	7.0	0.09385	12.1	−0.03716
2.0	0.45465	7.1	0.10267	12.2	−0.02936
2.1	0.41105	7.2	0.11023	12.3	−0.02140
2.2	0.36750	7.3	0.11650	12.4	−0.01336
2.3	0.31422	7.4	0.12145	12.5	−0.00531
2.4	0.28144	7.5	0.125072	12.6	0.00267
2.5	0.12939	7.6	0.12736	12.7	0.01049
2.6	0.19827	7.7	0.12833	12.8	0.01809
2.7	0.15829	7.8	0.12802	12.9	0.02539
2.8	0.11964	7.9	0.12645	13.0	0.03232
2.9	0.08250	8.0	0.12367	13.1	0.03883
3.0	0.04704	8.1	0.11974	13.2	0.04485
3.1	0.01341	8.2	0.11472	13.3	0.05034
3.2	−0.01824	8.3	0.10870	13.4	0.05525
3.3	−0.04780	8.4	0.10174	13.5	0.05954
3.4	−0.07516	8.5	0.09394	13.6	0.06317
3.5	−0.10022	8.6	0.08540	13.7	0.06613
3.6	−0.12292	8.7	0.07620	13.8	0.06838
3.7	−0.14320	8.8	0.06647	13.9	0.06993
3.8	−0.16101	8.9	0.05629	14.0	0.07076
3.9	−0.17635	9.0	0.04579	14.1	0.07087
4.0	−0.78920	9.1	0.03507	14.2	0.07028
4.1	−0.19958	9.2	0.02423	14.3	0.06901
4.2	−0.20752	9.3	0.01338	14.4	0.06706
4.3	−0.21306	9.4	−0.00264	14.5	0.06448
4.4	−0.21627	9.5	−0.00791	14.6	0.06129
4.5	−0.21723	9.6	−0.01816	14.7	0.05753
4.6	−0.21602	9.7	−0.02802	14.8	0.05326
4.7	−0.21275	9.8	−0.03740	14.9	0.04852
4.8	−0.20753	9.9	−0.04622	15.0	0.04335
4.9	−0.20050	10.0	−0.05440		
5.0	−0.19179	10.1	−0.06789		

Table F.2
Significant Values of the Radiation Pattern

x	$\sin\theta = x\lambda/\pi D$	θ	Amplitude $F(\theta)$
0	0	0	1
3.14	λ/D	Arc sin λ/D	0
4.5	1.43 λ/D	Arc sin 1.43 λ/D	0.21723
6.28	6.28 λ/D	Arc sin 2 λ/D	0
7.7	2.45 λ/D	Arc sin 2.45 λ/D	0.12833
9.45	3 λ/D	Arc sin 3 λ/D	0

If we now rotate the vertical plane around the segment, the pattern does not change. The diagram in the horizontal plane is a circle.

In conclusion, this result has been the subject of the following theorem: a series of point sources in phase balance placed on a segment of length D see their radiation concentrated in a beam of approximately a half angle α such that:

$$\sin\alpha = \lambda/D$$

which characterizes this network directivity of point sources.

We thus see that the longer D will be before the wavelength, the smaller α will be and therefore the narrower the beam will be. As N is proportional

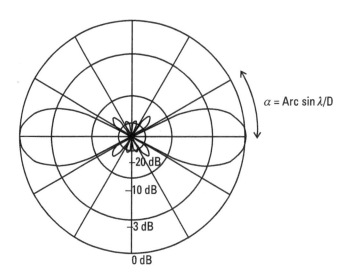

α = Arc sin λ/D

Figure F.6   Radiation pattern: vertical plane.

to D and the gain is proportional to N, we can say that the greater the gain will be, the narrower the beam will be.

Reference

[1]   Fouillé, A., *Physique des Vibrations* (in French), Dunod, 1954.

Appendix G

Transmission Line In High Frequency Reminders

G.1 Definition

A transmission line consists of two parallel conductive wires at distance d and separated by a dielectric (Figure G.1).

Let l be the length of a line, powered by an HF generator (G) and loaded at the other end by a load $Z_R = R + jX$.

G.2 Equations of Propagation Along the Line

In high frequency the line can be represented by a series of quadripoles in cascade (Figure G.2).

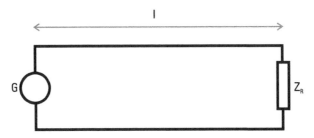

Figure G.1   Two-wire line.

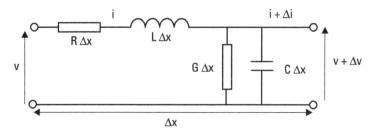

Figure G.2 Δx Element of the line.

R is the linear resistance per unit length.
L is the linear inductor per unit length.
G is the parallel conductance per unit length.
C is the parallel capacitance per unit length.

We can write:

$$\Delta v = i(R\Delta x) + (L\Delta x)\, di/dt \tag{G.1}$$

$$\Delta i = (v + \Delta v)\, (G\Delta x) + (C\Delta x)\, dv/dt = v\, (G\Delta x) + (C\Delta x)\, dv/dt \tag{G.2}$$

We can write these equations as the following:

$$dv/dx = Ri + Ldi/dt \tag{G.3}$$

$$di/dx = Gv + Cdv/dt \tag{G.4}$$

Equations (G.3) and (G.4) are difficult to solve in the general case.

We can easily see the shape of the solution for a line without loss ($R = 0$, $G = 0$). In that case:

$$dv/dx = Ldi/dt$$

$$di/dx = Cdv/dt$$

By differentiating with respect to x:

$$d^2v/dx^2 = L\, d/dx\, (di/dt) = Ld/dt\, (di/dx) = LC\, d^2v/dt^2$$

As a consequence:

$$d^2v/dx^2 = LC\, d^2v/dt^2 \tag{G.5}$$

$$d^2i/dx^2 = LC\, d^2i/dt^2 \tag{G.6}$$

These equations show that the voltage and current obey the wave differential equation of the second order seen in Appendix F. This equation is written as

$$d^2u/dz^2 - 1/c^2 d^2u/dt^2 = 0 \tag{G.7}$$

where u is the instantaneous amplitude of the wave, c its propagation velocity, and oz the propagation direction. The general solutions of this equation are of the form:

$$\bar{u}(z,t) = f_1(t - z/c) + f_2(t + z/c)$$

As we have seen in Appendix F, the term f_1 describes a wave propagating in the positive direction of the z and the term f_2 describes a wave propagating in the opposite direction. In the case of $f_2 = 0$ the wave is said to be a *traveling wave*. In the case where we have f_1 and f_2, we have a *direct wave* and a *reflected wave*.

If we compare (G.5) and (G.6) with (G.7), we can write:

$$c = 1/\sqrt{LC}$$

The current i and the voltage v form, thus waves propagate along the line without loss with a speed of:

$$c = 1/\sqrt{LC}$$

The wavelength is $\lambda = cT$ (see Appendix F). By solving (G.5) and (G.6), we obtain the propagation equations of the voltage and current on the line:

$$v = A\sin(\omega t - \beta x) + B\sin(\omega t + \beta x) \tag{G.8}$$

$$i = A/Z_0 \sin(\omega t - \beta x) - B/Z_0 \sin(\omega t + \beta x) \tag{G.9}$$

where $\beta = \omega\sqrt{LC}$ and $Z_0 = \sqrt{LC}$.

A and B are constants. They can be real or imaginary. We can note that we find:

- The direct wave $A \sin(\omega t - \beta x)$ for the voltage and $A/Z_0 \sin(\omega t - \beta x)$ for the current.
- The reflected wave $B \sin(\omega t + \beta x)$ for the voltage and $-B/Z_0 \sin(\omega t + \beta x)$ for the current.

These waves are indeed of the form $f_1(t - x/c)$ and $f_2(t + x/c)$ since $\beta = \omega/c$. The equation of the current traveling wave, for example, is

$$i = A/Z_0 \sin(\omega t - \beta x)$$

which is represented in Figure G.3

G.3 Interpretation of the Solution

- The quantity $Z_0 = \sqrt{L/C}$ has the dimension of an impedance and is expressed in ohms. This is called the *characteristic impedance* of the line. In lossless lines Z_0 is a pure resistance.

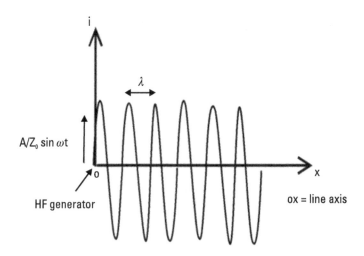

Figure G.3   Current traveling wave at time *t*.

- The quantity $\beta = \omega\sqrt{LC}$ is the *phase constant* seen in Chapter 6. Indeed, β can be written ω/c, $\lambda = cT$ and $T = 2\pi/\omega$.

Therefore, $\beta = 2\pi/\lambda = $ *phase constant* (see Appendix F). β is expressed in radian/meter. This means that, after traveling 1m, the wave has undergone a phase shift (for the voltage and current) of β radians.

G.3.0.1 Examples of application of the phase constant β.

Question: A line of length $\lambda/2$ is supplied by an HF generator. What is the output phase shift?

Answer: $(2\pi/\lambda) \times \lambda/2 = n$. The output voltage is in phase opposition with the input voltage.

Question: A line of length 11m is supplied by a 150-MHz frequency generator. What is the output phase shift?

Answer: $\lambda = c/f = 2\text{m}$.

$$(2\pi/\lambda)x11 = 11\pi = \pi$$

The output voltage is in phase opposition with the input voltage.

G.4 Reflection Coefficient

In the preceding paragraphs the origin of x was taken from the generator.

Let us take it now on the load Z_R. We just have to replace x with $-x$. At any point of the line we have:

$$v_{\text{dir}} = A \sin (\omega t + \beta x)$$

$$v_{\text{ref}} = B \sin (\omega t - \beta x)$$

We can write (see Appendix C):

$$\overline{v}_{\text{dir}} = A e^{j\omega t} e^{j\beta x}$$

$$\overline{v}_{\text{ref}} = B e^{j\omega t} e^{-j\beta x}$$

The reflection coefficient at the point of abscissa x is the ratio:

$$\overline{\Gamma}_x = \overline{v}_{\text{ref}}/\overline{v}_{\text{dir}} \qquad (G.10)$$

So: $\overline{\Gamma}_x = B/Ae^{-2j\beta x}$

We can see B/A is the reflection coefficient $\overline{\Gamma}_0$ on the charge Z_R because $x = 0$. Therefore:

$$\overline{\Gamma}_x = \overline{\Gamma}_0 e^{-2j\beta x} \tag{G.11}$$

G.5 Line Impedance

An impedance at a point of a line is the ratio:

$$\overline{Z}_x = \overline{v}(x)/\overline{i}(x) = (\overline{v}_{dir} + \overline{v}_{ref})/(\overline{i}_{dir} + \overline{i}_{ref})$$

Let

$$\overline{Z}_x = Z_0(Ae^{j\beta x} + Be^{-j\beta x})/(Ae^{j\beta x} - Be^{-j\beta x})$$
$$= Z_0(1 + B/A\ e^{-2j\beta x})/(1 - B/A\ e^{-2j\beta x})$$

Hence

$$\overline{Z}_x = Z_0(1 + \overline{\Gamma}_x)/(1 - \overline{\Gamma}_x) \tag{G.12}$$

for $x = 0$, we have

$$\overline{Z}_R = Z_0(1 + \Gamma_0)/(1 - \Gamma_0) \tag{G.13}$$

Inversely, we can write:

$$\overline{\Gamma}_0 = (\overline{Z}_R - Z_0)/(\overline{Z}_R + Z_0) \tag{G.14}$$

The equation of the line impedance can also be written as

$$\overline{Z}_x = Z_0(e^{j\beta x} + B/Ae^{-j\beta x})/(e^{j\beta x} - B/Ae^{-j\beta x})$$

Replacing B/A by $(\overline{Z}_R - Z_0)/(\overline{Z}_R + Z_0)$ we find:

$$\bar{Z}_x = Z_0[\bar{Z}_R(e^{j\beta x} + e^{-j\beta x}) + Z_0(e^{j\beta x} - e^{-j\beta x})]/[Z_0(e^{j\beta x} + e^{-j\beta x}) + \bar{Z}_R(e^{j\beta x} - e^{-j\beta x})]$$

By application of the following formulas (see Appendix C):

$$\cos\alpha = (e^{j\alpha} + e^{-j\alpha})/2$$
$$\sin\alpha = (e^{j\alpha} - e^{-j\alpha})/2j$$

we find

$$\bar{Z}_x = \bar{Z}_0(\bar{Z}_R\cos\beta x + jZ_0\sin\beta x)/(Z_0\cos\beta x + j\bar{Z}_R\sin\beta x)$$

and

$$\bar{Z}_x = \bar{Z}_0(\bar{Z}_R + jZ_0tg\beta x)/(Z_0 + j\bar{Z}_R tg\beta x) \qquad (G.15)$$

This is the equation of the impedance Z_x at a point of a line loaded by an impedance Z_R, for a line without loss.

G.5.1 Application 1

Input impedance of a closed line on a short-circuit. By using (G.15), we can calculate the impedance $Z(x)$ of a line in short circuit:

$$\bar{Z}_x = Z_0tg\beta x$$

We can plot Figure G.4, the curve of the $Z(x)$ module, as a function of x, and we can deduce that

- For $x < \lambda/4$, the line acts as an inductor.
- For $x = \lambda/4$, the line is equivalent to a resonant parallel circuit at resonance ($Zx = $ infini).
- For $\lambda/4 < x < \lambda/2$, the line behaves like a capacitor.
- For $x = \lambda/2$, the line is equivalent to a series resonant circuit at resonance ($Zx = 0$).

The line $\lambda/4$ short-circuited is the basic element of resonant coaxial cavities, widely used in filters in the VHF and UHF bands due to its property of being equivalent to a parallel resonant circuit.

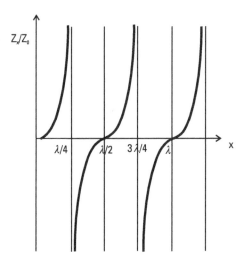

Figure G.4   Input impedance of a line in short circuit.

G.5.2 Application 2

Calculate the input impedance of a line $\lambda/4$ without loss loaded by a real impedance Z_R (pure resistance) with a characteristic impedance Z_0.

$$Z_x = Z_0(1 + \overline{\Gamma}_0 e^{-2j\beta x})/(1 - \overline{\Gamma}_0 e^{-2j\beta x})$$

For $x = \lambda/4$, $e^{-2j\beta x} = e^{-j\pi} = -1$ because $e^{-j\alpha} = \cos\alpha - j\sin\alpha$. Therefore

$$\overline{Z}_x = Z_0(1 - \overline{\Gamma}_0)/(1 + \overline{\Gamma}_0)$$

According to (G.14), Γ_0 is real because Z_R is real.

$$\Gamma_0 = (Z_R - Z_0)/(Z_R + Z_0) \tag{G.16}$$

$$Z_x = Z_0(Z_R + Z_0 - Z_R + Z_0)/(Z_R + Z_0 + Z_R - Z_0) = Z_0^2/Z_R$$
$$Z_x = Z_0^2/Z_R \tag{G.17}$$

G.6 Line Terminated by its Characteristic Impedance

In this very important case in practice (Figure G.5), $Z_R = Z_0$.

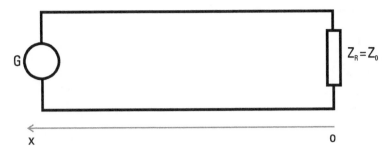

Figure G.5 Matched line.

The reflection coefficient Γ_0 is zero on the load [by (G.14)] and any point on the line $\Gamma_x = 0$ [from (G.15)]. There is no reflected wave. All the carried power is transmitted to the load. There is, as was said, a matching of the load to the line. Thus, we have a traveling wave.

G.7 Mismatched Line

G.7.1 General Case

Let there be a line with a characteristic impedance Z_0 loaded by any impedance $\bar{Z}_R = R + jX$ (Figure G.6). As the load impedance is complex, the reflection coefficient of the load is necessarily complex.

We can thus write:

$$\bar{\Gamma}_0 = \Gamma_0 e^{-j\Phi_0}$$

The reflection coefficient at a point of abscissa x (G.11) becomes

$$\bar{\Gamma}_x = \Gamma_0 e^{-j\Phi_0} e^{-2j\beta x} = \Gamma_0 e^{-j(\Phi_0 + 2j\beta x)} \tag{G.18}$$

In a line without losses, we have thus the relations:

$$\Gamma_x = \Gamma_0$$

$$\Phi = \Phi_0 + 2\beta x$$

Hence $\bar{\Gamma}_x$ in the complex plane (Figure G.7) where the arrow direction indicates the growth of the phase delay.

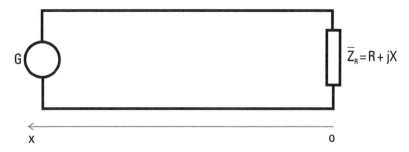

Figure G.6   Mismatched line.

We can deduce from this diagram that when the load is mismatched, we obtain a reflected wave whose reflection coefficient is constant at each point of a line without losses and equal to the reflection coefficient of the load.

We have by definition (G.8):

$$\overline{\Gamma}_x = \overline{v}_{\text{ref}} / \overline{v}_{\text{dir}}$$

Hence from (G.17):

$$\overline{v}_{\text{ref}} = \Gamma_0 \overline{v}_{\text{dir}} e^{-j(\Phi_0 + 2j\beta x)}$$

The total voltage at a point of the line is: $\overline{v} = \overline{v}_{dir} + \overline{v}_{ref}$. We can thus write:

$$\overline{v} = \overline{v}_{\text{dir}}(1 + \Gamma_0 e^{-j(\Phi_0 + 2j\beta x)})$$

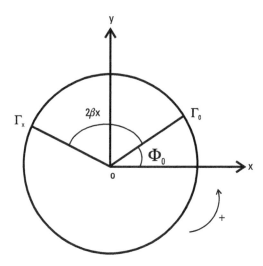

Figure G.7 Γ_x in the complex plane.

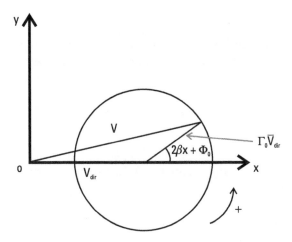

Figure G.8   Representation of $\bar{V}$.

This equation allows us to represent v in the complex plane from the previous Figure G.7.

We can deduce from this diagram that when we move on the line (x variable), the modulus V passes through a maximum and a minimum:

$$V_{max} = V_{dir}(1 + \Gamma_0)$$

$$V_{min} = V_{dir}(1 - \Gamma_0)$$

This diagram also allows us to draw the resultant wave directed to the load at a given time (Figure G.9).

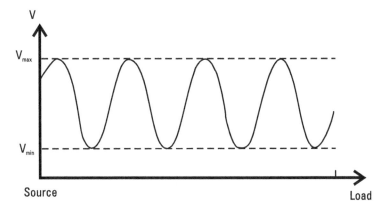

Figure G.9   Resultant wave at a given time.

G.7.2 Open-Line

In the case of an open-line, $\Gamma_0 = 1$ and $\Phi_0 = \pi$. All the power comes back to the source. We have a total reflection and the line acts as a pure reactance. The voltage $\bar{v}$ becomes

$$\bar{v} = \bar{v}_{\mathrm{dir}}\left(1 + e^{-2j\beta x}\right)$$

which can be represented in the complex plane (Figure G.10).

We can deduce from Figure G.10 the variation of the modulus V as a function of x:

V is a maximum and equal to $2V_{\mathrm{dir}}$ for $2\beta x = 2k\pi$, thus $x = k\lambda/2$;

V is zero for $2\beta x = (2k+1)\pi$, thus $x = (2k+1)\lambda/4$.

These are the characteristics of a standing wave (see Appendix F). We can therefore draw the standing wave (Figure G.11).

We have represented the voltage along the line. The current would have the same standing wave form, with a maximum current of V_{max}/Z_0 and a phase shift of $\lambda/4$, since on the load the current is zero (open circuit) and the voltage is maximum.

It is important to note that each wire of an open line fed by an HF source is traversed by a direct voltage wave and a reflected voltage wave of the same amplitude, but out of phase of π. The result is that on each wire we obtain a standing wave.

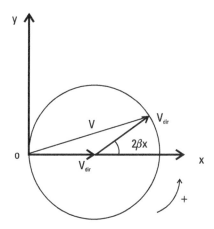

Figure G.10 V in the complex plane.

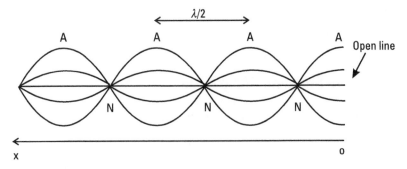

Figure G.11 Standing wave open line.

The standing wave voltage measured between the two wires of the line has a maximum amplitude equal to $2V_{dir}$ (see above), therefore the voltage standing waves on each wire are in phase opposition and have a maximum amplitude equal to V_{dir}.

We could say the same thing for the current, and thus we can write the following two properties of an open line:

1. The voltages and currents in each wire of the line, at the same distance from the HF source, are in phase opposition.

2. If the two wires of the line present a spacing much smaller than λ ($d = \lambda$), the electric and magnetic fields created by the currents of the two wires at a point M will also be in phase opposition with the same amplitude. In these conditions an open line does not radiate.

G.7.3 Line Closed by a Short Circuit

In the case of a line closed by a short circuit, we have a total reflection, and the line acts as pure reactance. We could demonstrate that we also obtain a

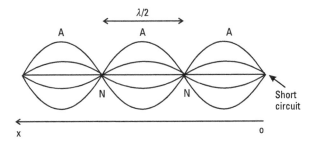

Figure G.12 Standing wave closed line by short circuit

standing wave for the voltage and current and that on the short circuit we have a voltage node and a current antinode (Figure G.12).

G.8 Voltage Standing Wave Ratio

The voltage standing wave radio is the ratio S:

$$S = V_{max}/V_{min}$$

as

$$V_{max} = V_{dir} + V_{ref}$$

and

$$V_{min} = V_{dir} - V_{ref}$$

we have

$$S = (1 + V_{ref}/V_{dir})/(1 * V_{ref}/V_{dir})$$

therefore

$$S = (1 + \Gamma)/(1 - \Gamma)$$

with Γ being the modulus of $\Gamma(x)$.

We can also write:

$$\Gamma = (S - 1)/(S + 1)$$

When the line is matched, $\Gamma = 0$ and $S = 1$.

G.9 Calculation of the Current in an Open Line without Loss of Length L

We can deduce from (G.8) and (G.9) (origin of x on the generator):

$$\bar{v} = (Ae^{-j\beta x} + Be^{j\beta x})e^{j\omega t}$$

With the Euler equations (Appendix C), we can write:

$$\bar{v} = [A(\cos\beta x - j\sin\beta x) + B(\cos\beta x + j\sin\beta x)]e^{j\omega t}$$

$$= [(A + B)\cos\beta x - j(A - B)\sin\beta x]e^{j\omega t}$$

$$\bar{i} = (A/Z_0 e^{*j\beta x} - B/Z_0 e^{j\beta x})e^{j\omega t}$$

$$= A/Z_0(\cos\beta x - j\sin\beta x) - B/Z_0(\cos\beta x + j\sin\beta x)e^{j\omega t}$$

$$= \left[\frac{(A - B)}{Z_0}\cos\beta x - j\frac{(A + B)}{Z_0}\sin\beta x\right]e^{j\omega t}$$

For $x = 0$ $\quad \bar{v} = v_0 e^{j\omega t}$; input voltage of the line

$$\bar{i} = I_0 e^{j\omega t}; \text{ input current of the line}$$

Thus:

$$V_0 = A + B$$

$$I_0 = (A - B)/Z_0$$

As a consequence:

$$\bar{i} = (I_0\cos\beta x - jV_0/Z_0\sin\beta x)e^{j\omega t}$$

For $x = L$ $\quad i = 0$, we can write:

$$0 = (I_0\cos\beta L - jV_0/Z_0\sin\beta L$$

Let $I_0 = jV_0/Z_0\sin\beta L/\cos\beta L$; hence, $\bar{i} = jV_0/Z_0\ [(\cos\beta x\ \sin\beta L - \sin\beta x\ \cos\beta L)/\cos\beta L]e^{j\omega t}$. The terms in parentheses are of the form $\sin(a - b)$. We thus find:

$$\bar{i} = jV_0/Z_0[\sin\beta(L - x)/\cos\beta L]e^{j\omega t} \tag{G.19}$$

This result can be simplified by taking the source of the current to the current I_M corresponding to the antinode closest the end of the line instead of I_0. I_M has as a value, according to (G.19):

$$I_M = jV_0/Z_0\ [\sin\beta(L - x_M) / \cos\beta L]$$

where x_M is the position of the current antinode.
as $(L - x_M) = \lambda/4$ and $\beta(L - x_M) = \pi/2$
Therefore:

$$\sin \beta(L - x_M) = 1$$

hence:

$$jV_0/Z_0 = I_M \cos \beta L$$

So the formula (G.17) becomes:

$$i = I_M \sin \beta(L - x) \qquad\qquad (G.20)$$

G.10   Types of Transmission Lines

Let us forget the optical fibers and present two types of transmission lines:

1. The two-wire line;
2. The coaxial line.

These lines are represented in Figures G.13 and G.14.

Figure G.13   Two-wire line.

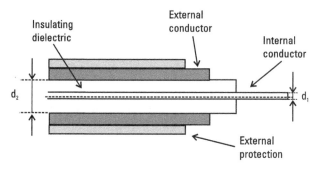

Figure G.14   Coaxial line.

σ is called the conductivity of the conductor.

ε is called the dielectric constant of the dielectric.

$tg\delta$ is called the angle of the dielectric ($tg\delta = C\omega/G$).

G.10.1 Characteristics of the Two-Wire Line

Let d be the diameter of the conductors, and D be their spacing from axis to axis, as indicated in Figure G.13.

The values of R, L, G, and C are given by the formulas:

R (ohm/m) $= (f/\pi\sigma)^{1/2} \, 1/d(1 - d^2/D^2)^{1/2}$

L (Henry/m) $= \mu/\pi \, \text{Log}(2D/d)$ (Log $=$ neperian logarithm)

G (Siemens/m) $= 2\pi^2\varepsilon f tg\delta/\text{Log}(2D/d)$

C (Farad/m) $= \pi\varepsilon/\text{Log}(2D/d)$

G.10.2 Characteristic of the Coaxial Line

Let: d_1 be the diameter of the inner conductor and d_2 be the inner diameter of the outer conductor (plait), as shown in Figure G.14.

The values of R, L, G, and C are given by the formulas:

R (ohm/m) $= (f/\pi)^{1/2} \, [1/d_1(\sigma_1)^{1/2} + 1/d_2 \, (\sigma)^{1/2}]$

L (Henry/m) $= \mu/\pi \, \text{Log}(d_2/d_1)$

G (Siemens/m) $= 4\pi^2\varepsilon f tg\delta/\text{Log}(d_2/d_1)$

C (Farad/m) $= 2\pi\varepsilon/\text{Log}(d_2/d_1)$

G.10.3 Comparison of the Two-Wire Lines and the Coaxial Lines

It follows from the previous formulas that the losses in the two-wire and coaxial lines increase:

- With the frequency.
- With the resistivity of the conductor ($\rho = 1/\sigma$).
- With the loss angle of the dielectric $tg\delta$.

The coaxial cables compared to the two-wire lines have a better conductivity and a much lower loss angle thanks to the use of a dielectric of polyethylene or Teflon. As a result, coaxial cables can be used for a given high frequency over much greater distances. For this reason coaxial cables are used between a transmitter and its antenna in the VHF and the UHF band.

G.11 Two-Wire Lines in Microelectronics

The two-wire lines are the basic element of circuits used in high-frequency microelectronics. For this there are two categories of line (Figure G.15): microstrip lines and slotted lines. These lines consist of metal strips bonded to a substrate.

G.11.1 Microstrip Line

A microstrip line is a substrate with a low dielectric constant ε, which is covered with metal on its lower part, with a metal strip on its upper part.

G.11.2 Triplate Line

A triplate line consists of two metal plates covering the lower face and the upper face of the substrate and a metal strip in the center of the dielectric.

G.11.3 Slot Line

A slot line is formed of two metal strips stuck on the same side of the substrate, separated by a slot.

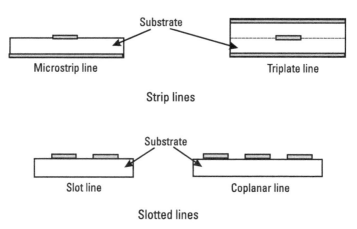

Figure G.15 Striplines and slotted lines.

G.11.4 Coplanar Line

In a coplanar line, three metal strips are stuck to the same side of the substrate and are separated by two slots. The central strip serves as a screen vis-a-vis the other two.

The most commonly used line is the microstrip line.

G.12 Applications of the Microstrip Line

Up to about 10 MHz, discrete components, resistors, inductors, and capacitors have characteristics independent from the frequency. It is the same for the conductive wires that connect these components together in a circuit. Here we are dealing with what is said to be elements with lumped constants.

Beyond about 10 MHz, these elements have resistors, inductors, and parasitic capacitances, so their characteristics have changed. It is the same for the wires that connect these components together in a circuit. Here we are dealing with what is said to be elements with distributed constants.

The technology used in microstrip lines, which are always very short relative to the wavelength (length in the order of $\lambda/10$), can be considered almost as elements with lumped constants. This is why it is possible to manufacture with these lines reliable components in the VHF and UHF bands and beyond: resistors, inductors, capacitors, and resonant circuits. Filters and matching circuits are produced in the VHF and UHF bands using these components. It is also possible to directly produce filters by coupling microstrip lines of length $\lambda/2$ and $\lambda/4$.

The coupling of microstip lines is also used for the manufacturing of directionnal couplers.

G.12.1 Production of Components

Let us consider a two-wire line, having a characteristic impedance Z_0, of length x terminated by an impedance Z_R (Figure G.16).

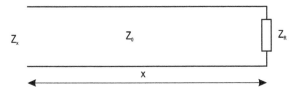

Figure G.16 Two-wire line.

The input impedance of this line is given by the formula seen in Section G.5, (G.15):

$$Z_x = Z_0(Z_R + jZ_0 tg\beta x)/(Z_0 + jZ_R tg\beta x)$$

If the microstrip lines are very short compared with the wavelength (x in the order of $\lambda/10$), we can write:

$$tg\beta x = \beta x = \pi/5$$

We then have

$$Z_x = Z_0(Z_R + jZ_0\pi/5)/(Z_0 + jZ_R\pi/5)$$
$$= Z_0(Z_R + jZ_0 0.63)/(Z_0 + jZ_R 0.63)$$

Let us assume that

$$Z_0 \ll Z_R$$

In this case, we can write:

$$Z_x = Z_0 Z_R/j0.63Z_R = -j1.59Z_0$$

The line is thus equivalent to a capacitor if its characteristic impedance is much lower than the load.

Let us now assume that

$$Z_0 \gg Z_R$$

In this case, we can write:

$$Z_x = jZ_0(0.63Z_0/Z_0) = j0.63Z_0$$

The line is thus equivalent to an inductor if its characteristic impedance is much larger than the load.

In the two cases the load Z_R can be replaced by another line of characteristic impedance Z_{01}. This is the principle used to make inductors and capacitors with microstrip lines (Figure G.17)

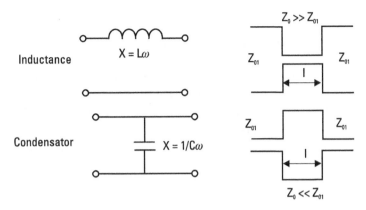

Figure G.17 Inductors and capacitors.

We have represented only the upper part of the substrate, the lower part being entirely covered with metal. We will not calculate the characteristic impedances. We will only retain that the characteristic impedance of a microstrip line is inversely proportional to its width.

We can see that the first pattern is indeed an inductor because the center line, having a smaller width than the two side lines, has a characteristic impedance greater than that of the two other lines. We can make a similar argument for the capacitor.

Figure G.18 shows a practical achievement on a 50 Ω line.

G.12.1.1 Conclusion: Microstrip-Line Components

Filters and matching circuits are made in the VHF and UHF bands with these components in cases when the required selectivity is not too high. These components are, for the moment, mainly used for making high-pass and low-pass filters in the VHF and UHF bands.

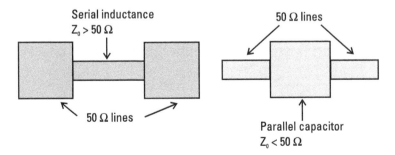

Figure G.18 Realization of a series inductor and a parallel capacitor on a 50 Ω line.

G.12.2 Directional Coupler

A directional coupler is another application of microstrip lines. It is based on the properties of two-coupled antennas, as seen in Chapter 4, Section 4.3.

Let us take two slightly coupled microstrip lines on the same substrate (Figure G.19). A transmitter is connected to the T_x input and the antenna output is connected to the antenna. Let us assume the antenna is mismatched, and the main line is traversed by two waves: a direct wave from the T_x input to the antenna output, and a reflected wave from the antenna output to the T_x input.

Each wave in the coupled line provides an electric wave in opposite direction. Therefore, a small part of the direct power kP_{dir} and reflected power kP_{ref} are distributed to the terminals of the coupled line, as indicated in Figure G.19. Thus, the directional coupler can be used to measure the direct power and the reflected power.

The opposite direction of each wave due to the coupling can be explained by the equations of two-coupled antennas as seen in Chapter 4, Section 4.3.

$$\bar{v}_1 = \bar{z}_{11}\dot{i}_1 + \bar{z}_{12}\dot{i}_2$$

$$\bar{v}_2 = \bar{z}_{21}\dot{i}_1 + \bar{z}_{22}\dot{i}_2$$

In our case $v_2 = 0$; therefore $i_2 = -i_1 Z_{21}/Z_{22}$.

A wave i_1 in the main line produces a wave i_2 out of phase of π in the coupled line; therefore the coupled line is traversed in the opposite direction.

G.12.3 Making Filters with Coupled Microstrip Lines

Another way to make filters is by using coupled microstrip lines, which are essentially bandpass filters. The properties of the lines $\lambda/2$ and $\lambda/4$ as resonators seen in Section G.5 are used. Thus if, on the same substrate, a microstrip line of characteristic impedance $50\,\Omega$ and length $\lambda/2$, is coupled on either side with other parallel lines of length $\lambda/2$, we obtain a bandpass filter.

G.13 A Short Review of the Smith Chart

Let's replace ox by oz. The Smith chart is an impedance chart that allows us to know $\Gamma(z)$ in order to calculate $Z(z)$, and vice versa for lossless lines. The relation between the reflection coefficient and the impedance at a point x of the line is given by (G.12):

$$Z_z = Z_0(1 + \Gamma_z)/(1 - \Gamma_z)$$

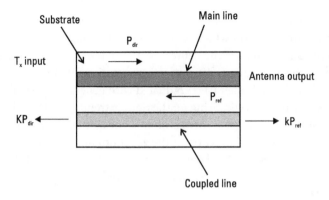

Figure G.19 Directional coupler.

which can be written by setting down

$$z_z = Z_z/Z_0$$

$$z_z = (1 + \Gamma_z)/(1 - \Gamma_z)$$

or

$$\Gamma_z = (1 + z_z)/(1 - z_z) \qquad (G.21)$$

We have on the other hand the relation (G.11):

$$\Gamma_z = \Gamma_0 \, e^{-2j\beta z}$$

By writing $-2j\beta z = \Phi$, we have $\Gamma_z = \Gamma_0 e^{j\Phi}$.

In the complex plane, the circles $\Gamma_z =$ constant are circles with center 0 and radius $\Gamma_0 <$ or $= 1$. Over the entire circle z varies from 0 to $\lambda/2$ (indeed for $\Phi = 2\pi$ we have $z = \lambda/2$).

If we set down $\Gamma_z = p + jq$ and $z_z = r + jx$, (G.21) will be written as:

$$\Gamma_z = (1 + r + jx) / (1 - r - jx)$$

By equalizing the real and imaginary parts, we obtain two equations representing r and x as a function of p and q:

$$[p - r/(1 + r)]^2 + q^2 = [1/(1 + r)]^2$$

$$(p - 1)^2 + (q - 1/x)^2 = (1/x)^2$$

These two equations represent the circles $r =$ constant and $x =$ constant shown in Figure G.21, on which we can draw the concentric circle $\Gamma =$ from $\Gamma = 0$ (zero radius circle at the center) to $\Gamma = 1$ (circle with radius equal to 1) according the application. The real axis x is positive, below it is negative.

The circle $\Gamma = 1$ (called Γ) is graduated with generally two scales z/λ and $\Phi = 2\pi x/\lambda$, the phase origin being on the right extremity of the real axis where the impedance is infinity (∞). There is often a linear scale for the measurement of the modulus of Γ. The direction of the generator and the load are generally indicated with the sign of Φ. Towards generator $\Phi > 0$, towards load $\Phi < 0$. The impedance $Z = R + jX$ is replaced by the reduced impedance $z = r + jx$ with $r = R/Z_0$ and $x = X/Z_0$.

The Smith chart has a lot of applications, impedance matching in particular, which is very usreful for antenna matching.

Let us present only this example whose application is made in the study of the Yagi antenna (Chapter 3).

G.13.3.1 Example

Let us assume we have a line of length 0.3m, with characteristic impedance 50 Ω, powered by a generator 50 Ω and loaded by an impedance $Z_R = 150 - j$ 100 Ω. The frequency is 300 MHz. Calculate with the Smith chart the module and the phase of the reflection coefficient at the input of the line.

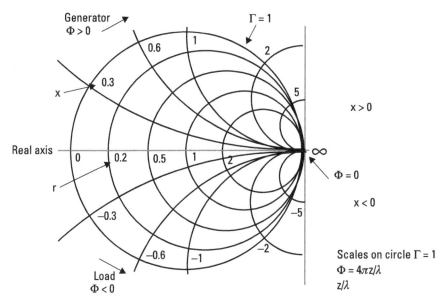

Figure G.20 Smith chart.

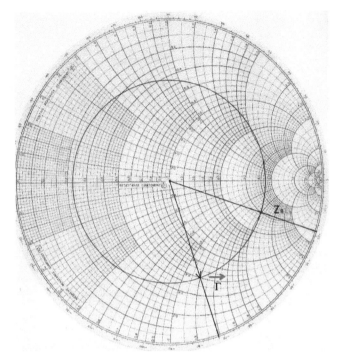

Figure G.21   Smith chart application example.

Solution (Figure G.21): The reduced impedance is $z = 1 - j1$. The point corresponding to this impedance is at the intersection of the circles $r = 1$ and $x = -1$. We can draw the circle Γ_z traversing this point.

The length of the line being 0.36m, the position of the generator on this circle is at a distance from the load of $z/\lambda = 0.36$, because $\lambda = 1$m. The scale is given on the circle $\Gamma = 1$ and for the load we have $z/\lambda = 0.34$. At this value, we have to add $z/\lambda = 0.36$ on the circle $\Gamma = 1$ in the direction indicated "Generator" (see Figure G.21). We find $z/\lambda = 0.20$. Thus the phase of the reflector coefficient Γ_s at the input of the line is positive and equal to:

$$\Phi_s = 4\pi(0.25 + 0.20) = 324°$$

we have to add the phase difference $2\pi \, l/\lambda = 54$ to the phase 18 of Z_R in the direction of the generator. Therefore the phase of Γ_x is 72 and its modulus is given by the ratio of both radius; that is to say: 0.63.

Its modulus is measured with the help of the vertical or horizontal scale: $|\Gamma_s| = 0.43$. We can conclude that the reflected voltage at the input of the line has a phase advance of 324° on the direct voltage.

References

[1]   Combes, P. F., *Micro-ondes* (in French), Dunod, 1996.

[2]   Septier, *Les Lignes de Transmissions* (in French), Paris University, 1956.

Appendix H

Wave Guide Reminders

H.1 Waveguides

Waveguides are hollow conductors, inside which electromagnetic waves are propagated by reflection on the inner walls. Waveguides are not fed like two-wire lines or coaxial cables by a generator or amplifier providing current and voltage, but by a small antenna or loop supplying an electric field or a magnetic field, respectively. There are two types of waveguides:

1. Rectangular waveguides;
2. Cylindrical waveguides.

We present only rectangular waveguides. To do this we must begin by:

1. Analyzing the reflection of plane electromagnetic waves on a conducting plane;
2. Analyzing the propagation between two parallel conducting planes;
3. Then we must add the third and fourth planes.

H.2 Reflection of a Plane Electromagnetic Wave on a Conductive Plane

Let us assume that we send an incident plane wave on a conducting plane with a propagation direction making an angle Ψ with the plane surface (Figure H.1). For each ray of the wave plane, the incidence plane is the plane formed by the propagation direction and by the normal n on the surface.

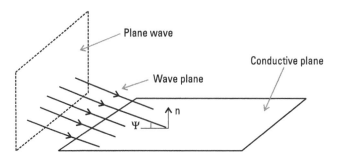

Figure H.1   Reflection on a conductive plane.

We will cover only the transverse electric (TE) plane waves.

We have a TE wave if the electric field is perpendicular to the incidence plane and the magnetic field is in the incidence plane.

If it's the contrary, we have a transverse magnetic (TM) wave. In this section, we are only discussing the TE wave (Figure H.2).

We assume that the reflection of the plane wave, of which we have represented only one ray, is on a vertical plane P. The incident electric field $\vec{E}_i$ has thus a vertical polarization.

We will use the following two properties of the good conductor (see Appendix D). The wave guide uses copper. At any point on its surface:

- The tangential component of the electric field $\vec{E}$ is zero; that is to say, the electric field is perpendicular to the surface;
- The normal component of the magnetic field $\vec{H}$ is zero; that is to say, the magnetic field is tangent to the surface.

The second characteristic property results from the electromagnetic wave:

- The magnetic field is perpendicular to the electric field (see Chapter 1).

At the impact point l, with the conductor being a good conductor, we must have:

$$\vec{E}_t \text{ and } \vec{H}_n = 0$$

$\vec{E}_i = 0$ is possible only if there is a reflected electric field equal and in phase opposition to the incident electric field. Hence $\vec{E}_r$, the direction of which is represented in Figure H.2.

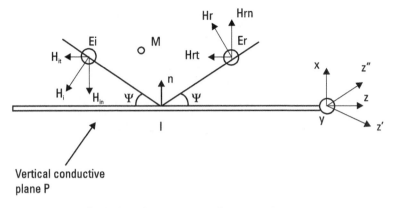

Figure H.2 TE wave (a single radius represented).

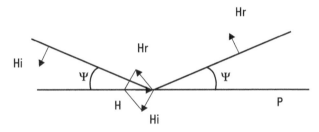

Figure H.3 Magnetic fields at the impact point.

$\vec{H}_n = 0$ means that there is a reflected field $\vec{H}_r$ at the impact point I symmetrical to $\vec{H}_i$ with respect to the surface of the conductor (Figure H.3). Therefore, there is a reflected wave at the same angle $\pi/2 - \Psi$.

The fields at a point M of the dielectric result from the superposition of the incident wave and of the reflected wave. To calculate the components of these two waves, we will take two additional axes:

1. oz', parallel to the direction of the incident wave.
2. oz'', parallel to the direction of the reflected wave.

The components of the fields $\vec{E}_i$ and $\vec{H}_i$ on the system of axes $oxyz$ can be written as:

$$\vec{E}_i \left| \begin{array}{l} E_{ix} = 0 \\[2mm] E_{iy} = -E_0 e^{j(\omega t - \beta z')} \\[2mm] E_{iz} = 0 \end{array} \right.$$

$$\vec{H}_i \left| \begin{array}{l} H_{ix} = -H_0 \cos \Psi e^{j(\omega t - \beta z')} \\ H_{iy} = 0 \\ H_{iz} = -H_0 \sin \Psi e^{j(\omega t - \beta z')} \end{array} \right.$$

with $\beta = 2\pi/\lambda$

The components of the fields $\vec{E}_r$ and $\vec{H}_r$ on the system of axes $oxyz$ can be written as:

$$\vec{E}_r \left| \begin{array}{l} E_{rx} = 0 \\ E_{ry} = -E_0 e^{j(\omega t - \beta z')} \\ E_{rz} = 0 \end{array} \right.$$

$$\vec{H}_r \left| \begin{array}{l} H_{rx} = -H_0 \cos \Psi e^{j(\omega t - \beta z')} \\ H_{ry} = 0 \\ H_{rz} = -H_0 \sin \Psi e^{j(\omega t - \beta z')} \end{array} \right.$$

Using the formulas of change of axes, we obtain, for the components of the fields $\vec{E}$ and $\vec{H}$ in M (see Figure H.2):

$$\vec{E} \left| \begin{array}{l} E_x = 0 \\ E_y = 2jE_0 e^{j\omega t} \sin(\beta x \sin \Psi) e^{-j\beta z \cos \Psi} \\ E_z = 0 \end{array} \right. \tag{H.1}$$

$$\vec{H} \left| \begin{array}{l} H_x = 2jH_0 e^{j\omega t} \cos \Psi \sin(\beta x \sin \Psi) e^{-j\beta z \cos \Psi} \\ H_y = 0 \\ H_z = -2H_0 e^{j\omega t} \sin \Psi \cos(\beta x \sin \Psi) e^{-j\beta z \cos \Psi} \end{array} \right.$$

We can calculate the energy in M by calculating the three components of the Poynting vector:

$$\vec{P} = \sqrt{\varepsilon\mu}\,\vec{E} \wedge \vec{H}$$

$$\vec{P} = \sqrt{\varepsilon\mu}\begin{vmatrix} \vec{i} & \vec{j} & \vec{k} \\ E_x & E_y & E_z \\ H_x & H_y & H_z \end{vmatrix}$$

That is, setting down $\sqrt{\varepsilon\mu} = k$:

$$\vec{P} = \begin{vmatrix} P_x = k(E_y H_z - E_z H_y) = -4kjE_0 H_0 \cos\Psi \sin(\beta y \sin\Psi)\cos(\beta y \sin\Psi) \\ P_y = k(E_z H_x - E_x H_z) = 0 \\ P_z = k(E_x H_y - E_y H_x) = 4kE_0 H_0 \cos\Psi \sin^2(\beta y \sin\Psi) \end{vmatrix}$$

The components of the Poynting vector show that

- There is no power conveyed in the direction Oy.
- The power propagating in the direction ox is a reactive power.
- The power propagating in the direction oz is an active power.

The propagation is thus characterized by the existence of

- A pure standing wave regime in a direction perpendicular to the surface of the conductor;
- A traveling wave regime in the direction oz.

H.3 Propagation of an Electromagnetic Plane Wave in the TE Mode between two Parallel Conductive Planes

H.3.1 Propagation Conditions

We will be able to introduce a plane P' parallel to P without interrupting the propagation only if the continuity conditions are satisfied at this level:

$E_1 = 0$ means $E_y = 0$ and $E_z = 0$
$H_n = 0$ means $H_x = 0$

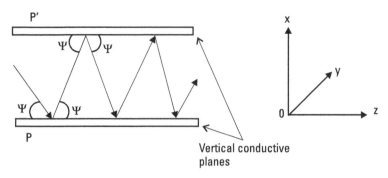

Figure H.4   Reflection on two planes.

Referring to field values of the combination of the incident wave and reflected wave at the point M, these conditions require:

$$\sin(\beta x \sin\Psi) = 0$$

Thus:

$$\beta x \sin\Psi = m\pi$$

If the distance between P and P′ is set to the value a, there can be propagation if

$$\sin\Psi = m\lambda/2a$$

with the condition

$$0 < m\lambda \quad 2a < 1$$

which defines a frequency band (λ) and modes (m).

H.4   Propagation of an Electromagnetic Plane Wave in the TE Mode in a Rectangular Waveguide

H.4.1   Propagation Conditions

We have just seen that it was possible, under certain conditions, to add a plane P′ parallel to P, and to obtain the propagation of a traveling wave in the direction *oz*. Let us now see if it is possible to add planes Q and Q' parallel to *xoz* and therefore perpendicular to P and P' without disrupting the propagation (Figure H.5).

The continuity conditions at their surface are

$$E_x = 0 \quad E_z = 0 \quad H_x = 0$$

Let Q and Q' be any two planes. We have necessarily on these two planes $E_x = 0$ $E_z = 0$ $H_x = 0$. It is therefore possible to add any two planes Q and Q'. The only condition is that

$$a = m\lambda/2 \sin\Psi$$

H.4.2 Definition of the TE$_{10}$ Mode

The TE wave propagation is carried out by successive reflections on the vertical planes P and P' perpendicular to *ox* and distant from

$$a = m\lambda/2 \sin\Psi \tag{H.2}$$

By convention, we will say that it is the TE$_{mo}$ mode. The excitation of this mode is made using a small rectilinear antenna (or probe) coupled to the transmitter and penetrating the guide vertically. If $m = 1$, the propagation in the waveguide is made according to the fundamental TE$_{10}$ mode, the most used in practice.

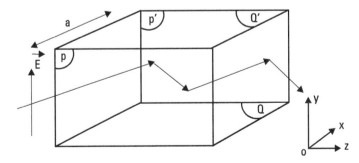

Figure H.5 Propagation in a rectangular guide.

H.4.3 Study of the Fundamental TE$_{10}$ Mode

The TE$_{10}$ mode is one that propagates inside the standard waveguide most commonly used in practice. It is said to operate on the fundamental mode. We are interested only in the electric field of the electromagnetic wave.

H.4.3.1 Calculation of the Electric Field at each Point within the Guide

By resuming (H.1) for the electric field components, we have:

$$\vec{E} = \begin{vmatrix} E_x = 0 \\ E_y = 2jE_0 e^{j\omega t} \sin(\beta x \sin\Psi) e^{-j\beta z \cos\Psi} \\ E_z = 0 \end{vmatrix} \qquad (H.3)$$

For $m = 1$ (H.2) becomes:

$$a = \lambda/2\sin\Psi$$

Let us replace β by $2\pi/\lambda$ and $\sin\Psi$ by $\lambda/2a$ in $\sin(\beta x \sin\Psi)$, we obtain:

$$\sin\beta x \sin\Psi = \sin \pi x/a$$

On the other hand, $\beta z \cos\Psi$, which is the phase term, shows that the wave length in the guide is

$$\lambda_g = \lambda/\cos\Psi$$

The electric field components become

$$\vec{E} = \begin{vmatrix} E_x = 0 \\ E_y = 2jE_0 \sin(\pi x/a) e^{j\omega t} e^{-j2\pi z/\lambda_g} \\ E_z = 0 \end{vmatrix} \qquad (H.4)$$

These equations show that in the TE$_{10}$ mode:

- The electric field at each point within the guide is parallel to 0y.
- Its amplitude is in $\sin \pi x/a$.

- The phase term $e^{-j2\pi z/\lambda_g}$ indicates that at each point, the electric field is presented in the form of a traveling wave propagating in the direction oz with a wavelength $\lambda_g = \lambda \cos\Psi$, which is greater than λ.

We have represented in Figure H.6 a wavelength of this traveling wave along a line PQ of the vertical median plane of the guide, on which the electric field amplitude is maximal: $2\,E_0$.

All the electric fields in the planes parallel to the plane xoy form an equiphase wave but not an equiamplitude wave.

H.4.3.2 Excitation of a Waveguide in TE_{10} Mode

The guide is excited by creating an electric field with a small antenna at the location where the field is maximum and therefore in the vertical median plane (Figure H.7).

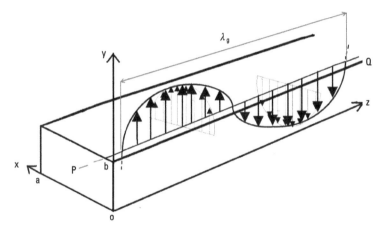

Figure H.6 Traveling wave in the TE_{10} mode.

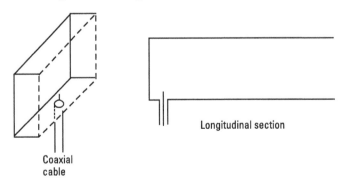

Figure H.7 Excitation of a waveguide in TE_{10} mode.

It is also possible to excite the guide with a small loop entering the guide at the same place. This loop provides a magnetic field with which a perpendicular electric field is associated, according to Maxwell's equations.

Reference

[1]   Combes, P. F., *Micro-ondes* (in French), Dunod, 1996.

About the Author

Bruno Delorme is an ESME Engineer and holds several university degrees in mathematics, electricity, and electronics. He has spent his entire career in radio communications, at TRT (12 years) as a transceiver design engineer at Motorola Communication (13 years), as a director of network engineering studies, and as a radio communication system consultant (10 years,) during which he conducted network studies and technical audits. He is also a trainer at the Training Centre of Radio-Data-Com (www.radio-data-com.fr).

Index

Marconi, Gugliemo, 7–8
Maxwell, James Clerk, 3
Maxwell-Ampere relation, 214
Maxwell equations
  electromagnetic waves, 9
  first, 205–6
  4th, 214
  list of, 10
  2nd, 213–14
  3rd, 214
  vertical half-wave dipole radiation
      through, 9–14
Maxwell-Faraday formula, 213
Maxwell theory, 3–4, 6
Mechanical tilt
  characteristics of, 103
  H-plane examples, 105
  illustrated, 104
  See also Tilt
Microstrip lines
  applications of, 247–50
  coupled, filters with, 250
  defined, 246
  directional coupler, 250
  production of components, 247–49
Microstrip-line technology
  coaxial cavity technology comparison,
      122–23
  defined, 120–21
Mismatched line, 237–42
  general case, 237–39
  illustrated, 238
  line closed by short circuit, 241–42
  open-line, 240–41
  resultant waves at a given time, 239
  total voltage, 238
  See also Transmission line
Multiband panel antennas, 111–12
Multicouplers
  advantages of, 129–30
  defined, 129
  receiving, 137–38
  transceiver, 142–43
  transmission, 130–37
  types of, 130
Multipaths
  defined, 109

  illustrated, 110
Mutual impedances, 90–93

N
Noise, measurement and protection,
      155–56
Notch filter duplexers
  defined, 139
  illustrated, 139
  industrial example, 148
  role of, 140
Notch filters
  industrial example, 126
  representation of, 117

O
Omnidirectional antennas
  E-plane, 20
  H-plane, 21
Omnidirectional vertical wire antennas,
      25–48
  defined, 30–31
  four-dipole, 43, 44–48
  ground plane, 36–37, 51
  half-wave, 32–33, 50
  industrial examples, 49–55
  infinitesimal dipole, 25, 26–29
  input impedance, 37–40
  introduction to, 25
  1.5 λ-length (three half-waves), 34–36
  λ-length (full wave), 33–34
  resonance and antiresonance, 40
  result analysis, 36–37
  six-dipole, 54
  two-dipole, 52
Open lines, 240–41, 242–44
Orthogonal polarization, 110–11
Oscillators
  current form in, 6
  spark gap, 5–6

P
Panel antennas, 55–65
  basic study, 56–61
  characteristic functions, 59, 60
  defined, 55
  dual band, industrial example, 115–16

Computational Electrodynamics: The Finite-Difference Time-Domain Method, Third Edition, Allen Taflove and Susan C. Hagness

Electromagnetic Modeling of Composite Metallic and Dielectric Structures, Branko M. Kolundzija and Antonije R. Djordjević

Electromagnetic Waves in Chiral and Bi-Isotropic Media, I. V. Lindell, et al.

Electromagnetics, Microwave Circuit and Antenna Design for Communications Engineering, Peter Russer

Engineering Applications of the Modulated Scatterer Technique, Jean-Charles Bolomey and Fred E. Gardiol

Fast and Efficient Algorithms in Computational Electromagnetics, Weng Cho Chew, et al., editors

Fresnel Zones in Wireless Links, Zone Plate Lenses and Antennas, Hristo D. Hristov

Handbook of Antennas for EMC, Thereza MacNamara

Handbook of Reflector Antennas and Feed Systems, Volume I: Theory and Design of Reflectors, Sudhakar Rao, Lotfollah Shafai, and Satish Sharma, editors

Handbook of Reflector Antennas and Feed Systems, Volume II: Feed Systems, Sudhakar Rao, Lotfollah Shafai, and Satish Sharma, editors

Handbook of Reflector Antennas and Feed Systems, Volume III: Applications of Reflectors, Sudhakar Rao, Lotfollah Shafai, and Satish Sharma, editors

Introduction to Antenna Analysis Using EM Simulators, Hiroaki Kogure, Yoshie Kogure, and James C. Rautio

Iterative and Self-Adaptive Finite-Elements in Electromagnetic Modeling, Magdalena Salazar-Palma, et al.

LONRS: Low-Noise Receiving Systems Performance and Analysis Toolkit, Charles T. Stelzried, Macgregor S. Reid, and Arthur J. Freiley

Measurement of Mobile Antenna Systems, Second Edition, Hiroyuki Arai

Microstrip Antenna Design Handbook, Ramesh Garg, et al.

Microwave and Millimeter-Wave Remote Sensing for Security Applications, Jeffrey A. Nanzer

Mobile Antenna Systems Handbook, Third Edition,
Kyohei Fujimoto, editor

Multiband Integrated Antennas for 4G Terminals,
David A. Sánchez-Hernández, editor

*Noise Temperature Theory and Applications for Deep Space
Communications Antenna Systems,* Tom Y. Otoshi

Phased Array Antenna Handbook, Second Edition,
Robert J. Mailloux

Phased Array Antennas with Optimized Element Patterns,
Sergei P. Skobelev

Plasma Antennas, Theodore Anderson

Quick Finite Elements for Electromagnetic Waves, Giuseppe Pelosi,
Roberto Coccioli, and Stefano Selleri

*Radiowave Propagation and Antennas for Personal
Communications, Second Edition,* Kazimierz Siwiak

Solid Dielectric Horn Antennas, Carlos Salema, Carlos Fernandes,
and Rama Kant Jha

Switched Parasitic Antennas for Cellular Communications,
David V. Thiel and Stephanie Smith

*Understanding Electromagnetic Scattering Using the Moment
Method: A Practical Approach,* Randy Bancroft

Wavelet Applications in Engineering Electromagnetics, Tapan
Sarkar, Magdalena Salazar Palma, and Michael C. Wicks

For further information on these and other Artech House titles, including previously considered out-of-print books now available through
our In-Print-Forever® (IPF®) program, contact:

Artech House
685 Canton Street
Norwood, MA 02062
Phone: 781-769-9750
Fax: 781-769-6334
e-mail: artech@artechhouse.com

Artech House
16 Sussex Street
London SW1V HRW UK
Phone: +44 (0)20 7596-8750
Fax: +44 (0)20 7630 0166
e-mail: artech-uk@artechhouse.com

Find us on the World Wide Web at: www.artechhouse.com